Peterson's®

egghead's Guide to Calculus

Cara Cantarella

About Peterson's®

Peterson's® is excited to be celebrating 50 years of trusted educational publishing. It's a milestone we're quite proud of, as we continue to provide the most accurate, dependable, high-quality education content in the field, providing you with everything you need to succeed. No matter where you are on your academic or professional path, you can rely on Peterson's publications and its online information at **www.petersons.com** for the most up-to-date education exploration data, expert test-prep tools, and the highest-quality career success resources—everything you need to achieve your educational goals.

For more information, contact Peterson's, 3 Columbia Circle, Albany, NY 12203-5158; 800-338-3282 Ext. 54229; or visit us online at **www.petersons.com**.

ISBN: 978-0-7689-3978-1

Printed in the United States

10 9 8 7 6 5 4 3 2 1   17 16 15

Contents

Contents

Contents

Contents

Before You Begin

Welcome to *egghead's Guide to Calculus*! My name is egghead, and I'll be your guide throughout the book.

Before You Begin

This egghead's Guide was designed to help you learn calculus in a fun and easy way. Sometimes learning can be . . . well, boring. It can also be confusing at times. If it wasn't, we'd all have straight A's, right?

As your guide through the adventure of education, I'm here to make things a bit more enjoyable. I studied the boring books so you don't have to. I got straight A's and lived to tell about it. I understand this stuff, and you can, too. In this guide, I'll show you what you need to learn to get to the next level.

Wherever I can, I explain things in pictures and stories. I break concepts down and teach them step by step. I try to stick with words that you know. I give examples from real life that you can relate to.

In this book, we'll work together to improve your calculus skills and build your confidence. Confidence is very important, and it comes from trust. You can trust me as your guide, and most important, you can trust yourself. If your calculus knowledge isn't strong enough, let's do something about it!

I want you to succeed, and I know you can do it!

How This Book Is Organized

This book contains nine chapters. We recommend you read them in order.

The introduction comes first.

Chapter 1 provides a review of algebra. In order to solve calculus problems, you'll need to use algebra often. This chapter reviews basic algebra concepts, such as multiplying and dividing polynomials, factoring polynomials, and absolute value. Here we'll also explain interval notation.

Chapter 2 discusses graphing. This chapter covers the coordinate plane, graphing linear equations, finding the slope of a line, the slope-intercept formula, and graphing quadratic equations.

Chapter 3 defines functions. Here we review the basic concept of a function and discuss domain and range. We also explain functional notation, composite functions, and piecewise functions.

Chapter 4 addresses trigonometry. Along with algebra, graphing, and functions, trigonometry is used throughout calculus. In this chapter, we review the Pythagorean Theorem and how to find sine, cosine, and tangent. We also address inverse trigonometric functions and the concept of the unit circle.

Chapter 5 introduces the concept of limits. Here we learn about the properties of limits and how to take limits of polynomials, rational functions, composite functions, and trig functions. Infinite limits are also addressed.

Chapter 6 deals with differentiation. This chapter introduces one of the two main contributions of calculus, the process of differentiation. We address how to understand differentiation and how to work with derivatives.

Chapter 7 focuses on applications of differentiation. The process of differentiation has many varied applications. This chapter presents several key applications and shows how to use differentiation to solve specific types of problems.

Chapter 8 explains integration. This chapter focuses on the second main contribution of calculus, the process of integration. We discuss how to understand integration and how to solve problems using integrals.

Chapter 9 concludes with definite integrals and the Fundamental Theorem of Calculus. The related processes of differentiation and integration are tied together by connections shown through the Fundamental Theorem of Calculus. This chapter explains the connections and shows how the Fundamental Theorem assists calculus problem-solving.

As you read each chapter, you'll find practice exercises along the way. Complete the exercises to practice what you've learned. The more you use your skills, the better they will "stick" in your mind.

To Learn More

Ready for more practice? After you've finished this book, visit the egghead website at www.petersonspublishing. com. Click the egghead link for additional practice exercises. This book will get you off to a great start. The website can give you that extra calculus boost!

Peterson's Books

Along with producing the egghead's Guides, Peterson's publishes many types of books. These can help you prepare for tests, choose a college, and plan your career. They can even help you obtain financial aid. Look for Peterson's books at your school's guidance office or library, your local library or bookstore, or at www.petersonsbooks.com. Many Peterson's eBooks are also online!

We welcome any comments or suggestions you may have about this book. Your feedback will help us make educational dreams possible for you—and others like you.

Introduction

The discipline of calculus was developed through the cumulative work of many mathematicians and scientists over hundreds of years. Before such key contributors as Sir Isaac Newton and Gottfried Leibniz, the scientific community was unable to solve many problems that had vexed scholars for ages. As the tools of calculus became more developed and refined, mathematicians were able to achieve solutions that had once been out of reach. With this mathematical advancement came a surge in technological progress, creating much of the capacity that we now have today.

Many of the problems that calculus helps to solve concern motion and change. We would use calculus to determine the instantaneous velocity of an object moving at a constantly changing speed. We would use calculus to determine the slope of a curved line or the area under a curve. Calculus can also be used to determine the length of a curved line, such as an arc. The engineering applications of calculus can be used to solve a myriad of mechanical and electrical problems, such as those involving the flow of fluids or the flow of electrical charge. Calculus further provides us with a new ability to understand and manipulate oscillating trigonometric functions.

To help you establish a solid foundation in calculus, this guide incorporates four teaching principles. First, the concepts of calculus are presented in the following different ways throughout the book:

- verbally, through descriptive explanations

- mathematically, by showing calculations

- visually, through graphical representations

- illustratively, through examples of concrete applications

These different approaches will help to bolster your understanding of the concepts by offering you multiple ways to comprehend the same ideas.

Second, the book begins with a refresher of math skills you need to know before you can master calculus. Four chapters are devoted to providing you with necessary background knowledge to use calculus problem-solving techniques. The chapters on algebra, graphing, functions, and trigonometry will bring you up to speed on any skills you might have forgotten (or never learned) that are required for completing later chapters in the book. If your background knowledge is strong, you can skip these chapters, or just consult them when you have a question about a particular skill. If your background knowledge is undeveloped, we recommend working through all of the concept reviews and exercises in the first four chapters before proceeding to the rest of the book.

A third teaching principle that defines the book is that it is written in easy-to-understand language that breaks the concepts down into manageable chunks with ample practice. Most of the content sections include practice exercises that address the same content presented in the teaching sections, reinforcing the content through repetition.

Before You Begin

Finally, this book focuses on the two most important contributions of calculus—differentiation and integration—in an in-depth way. The goal is for you to understand a smaller number of essential concepts very well, rather than presenting many concepts superficially. Once the basic building blocks of your knowledge are established, you will be able to approach more complex problems from a position of strength. Many great scholastic minds have contributed to inventing calculus over the years, building on the advances of those before them, step by step. With the benefit of this knowledge at your disposal, who knows what you might go on to contribute some day!

Chapter 1

Algebra Review

Hi! I'm egghead. In this chapter, we'll review the following concepts:

Multiplying and dividing polynomials
Factoring polynomials
Absolute value
Interval notation

Multiplying and dividing polynomials

Some calculus problems require you to work with polynomials. A **polynomial** is the sum of expressions containing variables and exponents. Polynomials can come in different forms. A **monomial** contains just one term, such as $3a^3b^2$. A **binomial** contains two unlike terms, such as $3a^3 + 3a$. A **trinomial** contains three unlike terms, such as $3a^3 + 3a + 5$, and so on.

Multiplying polynomials

When adding and subtracting polynomials, we combine like terms. **Like terms** are terms that have the same variable raised to the same power. To add the polynomials $(7x^2 + 4x + 2) + (3x^2 + 2x + 1)$, for instance, we would combine the like terms $7x^2 + 3x^2$, $4x + 2x$, and $2 + 1$ for a result of $10x^2 + 6x + 3$.

To multiply polynomials, we must work with the rules of exponents. Here is a chart summarizing the multiplication rules:

Expression	Operation	Result
$n^a \times n^b$	Add exponents	n^{a+b}
$(n^a)^b$	Multiply exponents	$n^{a \times b}$
$(n \times m)^a$	Raise bases to that power	$n^a \times m^a$

As the first rules shows, to multiply exponential expressions with the same base, you must add the exponents:

$$5y^2(y^2 + 7) = 5y^4 + 35y^2$$

To multiply $5y^2$ times y^2, we multiplied the coefficients 5×1 and added the exponents $2 + 2$. The result is $5y^4$.

When multiplying two binomials, we use the FOIL technique.

To multiply the binomials $(4a + 2)$ and $(a + 7)$, we would start by multiplying the first two terms, $4a \times a$. Next, we would multiply the two outer terms, $4a \times 7$. Then we would multiply the two inner terms, $2 \times a$, and finally we would multiply the last terms, 2×7.

FOIL stands for **F**irst, **O**uter, **I**nner, **L**ast.

Adding up the results, we would have the following:

$$4a^2 + 28a + 2a + 14$$

In the last step, we would simplify:

$$4a^2 + 30a + 14$$

Dividing polynomials

Dividing by polynomials also requires working with the rules of exponents. Here is a chart summarizing the division rules:

Expression	Operation	Result
$\dfrac{n^a}{n^b}$	Subtract exponents	n^{a-b}
$\left(\dfrac{n}{m}\right)^a$	Raise bases to that power	$\dfrac{n^a}{m^a}$
n^{-a}	Take reciprocal of base with positive exponent	$\dfrac{1}{n^a}$

To divide exponents with the same base, as the first rule shows, we must subtract exponents:

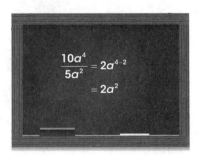

$$\frac{10a^4}{5a^2} = 2a^{4-2}$$
$$= 2a^2$$

To divide a polynomial by a monomial, we must treat the fraction as if it were multiple fractions and perform the division separately for each individual fraction:

$$\frac{4z^3 + 12z^2}{2z} = \frac{4z^3}{2z} + \frac{12z^2}{2z}$$
$$= 2z^{3-1} + 6z^{2-1}$$
$$= 2z^2 + 6z$$

The correct answer is $2z^2 + 6z$.

Practice Questions—Multiplying and dividing polynomials

Directions: Simplify the polynomials using multiplication or division. You will find the Practice Question Solutions on page 15.

1. $2c^3(9c^2 + 3)$

2. $4x(x^3 + 2x^2 + 1)$

3. $\dfrac{10a^2 + 8a}{2a}$

4. $\dfrac{100r^3 + 20r^2}{10r^2}$

5. $(x + 5)(x + 4)$

6. $(s + 7)(s + 3)$

7. $(n + 1)(n + 4)$

8. $(2p^2 + 2p)(p^2 + 5p + 4)$

9. $(3a^2 + 1)(5a^2 + 2a - 1)$

10. $(z + 6)^3$

Factoring polynomials

Certain calculus computations require separating a polynomial expression into its factors. Here are some examples showing how to factor.

Factoring binomials

To factor a binomial such as $5x^2 + 5$, we first identify the Greatest Common Factor, or GCF. The **Greatest Common Factor** of a polynomial is the largest factor that is included in all terms of the polynomial.

The greatest common factor of the binomial $5x^2 + 5$ is the number 5. The number 5 is a factor of both $5x^2$ and 5.

Separate the binomial into the product of two factors:

$$5x^2 + 5 = 5(x^2 + 1)$$

The factors are 5 and $(x^2 + 1)$.

Factoring trinomials

To factor a trinomial, we use reverse FOIL. To factor the trinomial $x^2 + 4x + 4$, set out the parentheses to be used for the two factors:

$$x^2 + 4x + 4 = (\quad)(\quad)$$

Insert the first variable, x, inside both parentheses:

$$x^2 + 4x + 4 = (x + \underline{})(x + \underline{})$$

The two remaining numbers must produce a product of 4 when multiplied together. Try the numbers 2 and 2:

$$x^2 + 4x + 4 = (x + 2)(x + 2)$$

The numbers 2 and 2 will work. To be sure, check your work using FOIL:

$$(x + 2)(x + 2) = x^2 + 2x + 2x + 4$$
$$= x^2 + 4x + 4$$

The factors are $(x + 2)$ and $(x + 2)$.

Practice Questions—Factoring polynomials

Directions: Factor the binomials and trinomials shown. You will find the Practice Question Solutions on page 17.

11. $2a^2 + 3a$

12. $z^2 + 11z$

13. $6d^2 - 6d$

14. $12b - 28$

15. $12r^2 - 108$

16. $q^2 + 6q - 16$

17. $4m^2 - 6m - 18$

18. $2x^2 + 16x - 40$

19. $y^2 - 11y + 18$

20. $n^2 - 196$

Absolute value

The **absolute value** of a number indicates how far the number lies from zero on a number line. The symbol for absolute value is two vertical bars:

$$|3| = 3$$

This equation shows that the absolute value of 3 is 3. This is correct, because the number 3 lies exactly 3 units away from zero on the number line.

It would also be correct to say that $|-3| = 3$. The number -3 lies exactly 3 units away from zero on the number line also, in the opposite direction.

As another example, consider the following. What is the value of n if the absolute value of n equals 7?

$$|n| = 7$$

In this case, the value of n could be 7 or -7. Both numbers lie exactly 7 units from zero on the number line. Therefore, the solutions are $n = 7$ or $n = -7$.

When solving absolute value equations, we must look for both the positive and negative values of the expression inside the bars.

$$|3 + y| = 7$$

In this case, the value of $3 + y$ could equal 7 or -7.

Set up two equations to determine both results:

$$
\begin{aligned}
|3 + y| &= 7 \\
3 + y &= 7 \\
y &= 7 - 3 \\
y &= 4
\end{aligned}
\qquad
\begin{aligned}
|3 + y| &= 7 \\
3 + y &= -7 \\
y &= -7 - 3 \\
y &= -10
\end{aligned}
$$

One possible value of y is 4, and the other possible value of y is 10.

The correct answer is $y = 4$ and $y = -10$.

Always solve for both possible solutions!

Practice Questions—Absolute value

Directions: Find the solutions to the absolute-value equations shown. You will find the Practice Question Solutions on page 17.

21. $|z + 7| = 9$

22. $|r - 3| = 14$

23. $|h + 2| = 20$

24. $|q - 7| = 6$

25. $|s + 15| = 30$

26. $|3m - 4| = 11$

27. $|t \div 7| = 4$

28. $|6a + 7| = 25$

29. $|4g \times 5| = 36$

30. $|14b \div 7| = 4$

Interval notation

An **interval** is a set of numbers that lie between two quantities. Say, for instance, we are interested in describing the set of numbers that are greater than 1 and less than 6. This interval could be expressed as follows:

This interval is **exclusive** because it excludes the numbers 1 and 6. It is also called an **open interval**, because it doesn't include its endpoints. If we drew the interval on a number line, it would look like this:

The open circles at points 1 and 6 show that the numbers 1 and 6 are excluded from the set.

Interval notation is a symbolic way of representing intervals. To represent an exclusive interval, we use parentheses ( ). The interval just described would be written in interval notation as (1, 6).

The parentheses tell us that the numbers 1 and 6 are **excluded.**

A set of numbers is called an **inclusive** set if the end points are contained within the set. Take, for example, the set that contains all numbers between 2 and 5, with 2 and 5 included. As an inequality, this set would be written as follows:

$$2 \le x \le 5$$

This inequality represents the set of all numbers greater than or equal to 2 and less than or equal to 5. It is also called a **closed interval** because it does include its endpoints.

On a number line, this inclusive interval would be drawn as follows:

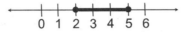

The shaded circles show that the numbers 2 and 5 are included in the set.

To express an inclusive interval in interval notation, we use brackets [ ]. The interval just described would be written as [2, 5].

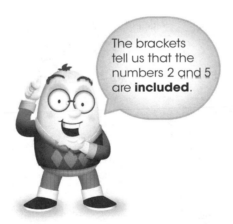

The brackets tell us that the numbers 2 and 5 are **included**.

It is possible to have an interval that includes one endpoint and excludes the other. These intervals are written with both brackets and parentheses. For instance, the set of all numbers greater than 1 and less than or equal to 5 would be written as (1, 5]. The parenthesis tells us that the number 1 is excluded, while the bracket tells us that the number 5 is included. As an inequality, this set would be expressed as: $1 < x \le 5$.

The set of all numbers greater than or equal to 2 and less than 6 would be written as [2, 6). The set of all numbers greater than 4 would be written as (4, ∞). The symbol ∞ is used to represent infinity. The set of all numbers is written as (−∞, ∞).

Practice Questions—Interval notation

Directions: Express the following intervals as inequalities. You will find the Practice Question Solutions on page 19.

31. $[3, 16]$

32. $(-5, 2)$

33. $[-11, -7)$

34. $(a, b]$

35. $[2s, s^2 - 3]$

Directions: Write the following inequalities using interval notation. You will find the Practice Question Solutions on page 19.

36. $0 < x < 12$

37. $-9 \leq x \leq 9$

38. $\frac{1}{2} < x \leq 21$

39. $a \leq x \leq 7b$

40. $\frac{a+2}{b} < x \leq \frac{a}{c}$

Chapter Review

Directions: Simplify the polynomials using multiplication or division. Solutions can be found on page 20.

1. $6a^3(4a^3 + 3)$

2. $\dfrac{28v^2 + 49}{7}$

3. $(c + 10)(c + 6)$

4. $(4s + 6)(9s - 7)$

5. $(7x^2 + 4)(6x^3 - 3x^2 + 2)$

Directions: Factor the binomials and trinomials shown. Solutions can be found on page 21.

6. $3t^2 + 7t$

7. $-h^2 + 5h$

8. $k^2 - 10k - 24$

9. $-a^2 + 6a - 9$

10. $2z^2 - 13z - 70$

Directions: Find the solutions to the absolute value equations shown. Solutions can be found on page 21.

11. $|3a - 12| = 3$

12. $|2z \times 9| = 48$

13. $|4d \div 16| = 4$

14. $|7b \times 3| = 42$

15. $|6j + 9| = 35$

Directions: Express the following as inequalities. Solutions can be found on page 22.

16. $(-2, 7)$

17. $[12, 23]$

18. $(-4, 2]$

19. $[y, 3z]$

20. $(6 + t, 2t^2 + 7]$

Directions: Write the following inequalities using interval notation. Solutions can be found on page 22.

21. $-1 \leq x < 7$

22. $6 < x \leq 14$

23. $\frac{3}{7} < x < 9$

24. $2c \leq x < 3d$

25. $2(a - 4) \leq x \leq \frac{a + b}{3}$

Practice Question Solutions

Multiplying and dividing polynomials

1. The correct answer is $18c^5 + 6c^3$.

    Distribute the $2c^3$ over the $9c^2$ and the 3:

    $$2c^3\left(9c^2 + 3\right) = \left(2c^3 \times 9c^2\right) + \left(2c^3 \times 3\right)$$
    $$= 18c^{3+2} + 6c^3$$
    $$= 18c^5 + 6c^3$$

2. The correct answer is $4x^4 + 8x^3 + 4x$.

    $$4x\left(x^3 + 2x^2 + 1\right) = \left(4x \times x^3\right) + \left(4x \times 2x^2\right) + \left(4x \times 1\right)$$
    $$= 4x^{1+3} + 8x^{1+2} + 4x$$
    $$= 4x^4 + 8x^3 + 4x$$

3. The correct answer is $5a + 4$.

    $$\frac{10a^2 + 8a}{2a} = \frac{10a^2}{2a} + \frac{8a}{2a}$$
    $$= 5a^{2-1} + 4a^{1-1}$$
    $$= 5a + 4$$

4. The correct answer is $10r + 2$.

    $$\frac{100r^3 + 20r^2}{10r^2} = \frac{100r^3}{10r^2} + \frac{20r^2}{10r^2}$$
    $$= 10r^{3-2} + 2r^{2-2}$$
    $$= 10r + 2$$

5. The correct answer is $x^2 + 9x + 20$.

$$(x+5)(x+4) = (x \times x) + (x \times 4) + (5 \times x) + (5 \times 4)$$
$$= x^2 + 4x + 5x + 20$$
$$= x^2 + 9x + 20$$

Use FOIL to multiply the binomials.

6. The correct answer is $s^2 + 10s + 21$.

$$(s+7)(s+3) = (s \times s) + (s \times 3) + (7 \times s) + (7 \times 3)$$
$$= s^2 + 3s + 7s + 21$$
$$= s^2 + 10s + 21$$

7. The correct answer is $n^2 + 5n + 4$.

$$(n+1)(n+4) = (n \times n) + (n \times 4) + (1 \times n) + (1 \times 4)$$
$$= n^2 + 4n + n + 4$$
$$= n^2 + 5n + 4$$

8. The correct answer is $2p^4 + 12p^3 + 18p^2 + 8p$.

$$(2p^2 + 2p)(p^2 + 5p + 4) = (2p^2 \times p^2) + (2p^2 \times 5p) + (2p^2 \times 4) + (2p \times p^2) + (2p \times 5p) + (2p \times 4)$$
$$= (2p^{2+2}) + (10p^{2+1}) + (8p^2) + (2p^{1+2}) + (10p^{1+1}) + (8p)$$
$$= 2p^4 + 10p^3 + 8p^2 + 2p^3 + 10p^2 + 8p$$
$$= 2p^4 + 10p^3 + 2p^3 + 8p^2 + 10p^2 + 8p$$
$$= 2p^4 + 12p^3 + 18p^2 + 8p$$

9. The correct answer is $15a^4 + 6a^3 + 2a^2 + 2a - 1$.

$$(3a^2 + 1)(5a^2 + 2a - 1) = (3a^2 \times 5a^2) + (3a^2 \times 2a) + [3a^2 \times (-1)] + (1 \times 5a^2) + (1 \times 2a) + [1 \times (-1)]$$
$$= (15a^{2+2}) + (6a^{2+1}) + (-3a^2) + (5a^2) + (2a) + (-1)$$
$$= 15a^4 + 6a^3 - 3a^2 + 5a^2 + 2a - 1$$
$$= 15a^4 + 6a^3 + 2a^2 + 2a - 1$$

10. The correct answer is $z^3 + 18z^2 + 108z + 216$.

$$(z+6)^3 = (z+6)(z+6)(z+6)$$

Multiply the first two binomials using FOIL:

$$(z+6)(z+6) = (z \times z) + (z \times 6) + (6 \times z) + (6 \times 6)$$
$$= z^2 + 6z + 6z + 36$$
$$= z^2 + 12z + 36$$

Using FOIL gives the trinomial $z^2 + 12z + 36$. Now, use distribution:

$$(z+6)(z^2+12z+36) = \left(z \times z^2\right) + (z \times 12z) + (z \times 36) + \left(6 \times z^2\right) + (6 \times 12z) + (6 \times 36)$$

$$= \left(z^{1+2}\right) + \left(12z^{1+1}\right) + (36z) + \left(6z^2\right) + (72z) + (216)$$

$$= z^3 + 12z^2 + 36z + 6z^2 + 72z + 216$$

$$= z^3 + 12z^2 + 6z^2 + 36z + 72z + 216$$

$$= z^3 + 18z^2 + 108z + 216$$

Factoring polynomials

11. The factors are a and $(2a + 3)$.

12. The factors are z and $(z + 11)$.

13. The factors are $6d$ and $(d - 1)$.

14. The factors are 4 and $(3b - 7)$.

15. The factors are 12, $(r + 3)$, and $(r - 3)$.

16. The factors are $(q - 2)$ and $(q + 8)$.

17. The factors are $(4m + 6)$ and $(m - 3)$.

18. The factors are $(2x - 4)$ and $(x + 10)$.

19. The factors are $(y - 9)$ and $(y - 2)$.

20. The factors are $(n + 14)$ and $(n - 14)$.

Absolute value

21. The solutions are $z = 2$ and $z = -16$.

$$z + 7 = 9 \qquad\qquad z + 7 = -9$$
$$z = 9 - 7 \qquad\qquad z = -9 - 7$$
$$z = 2 \qquad\qquad z = -16$$

22. The solutions are $r = 17$ and $r = -11$.

$$r - 3 = 14 \qquad\qquad r - 3 = -14$$
$$r = 14 + 3 \qquad\qquad r = -14 + 3$$
$$r = 17 \qquad\qquad r = -11$$

23. The solutions are $h = 18$ and $h = -22$.

$$h + 2 = 20 \qquad\qquad h + 2 = -20$$
$$h = 20 - 2 \qquad\qquad h = -20 - 2$$
$$h = 18 \qquad\qquad h = -22$$

24. The solutions are $q = 13$ and $q = 1$.

$$q - 7 = 6 \qquad\qquad q - 7 = -6$$
$$q = 6 + 7 \qquad\qquad q = -6 + 7$$
$$q = 13 \qquad\qquad q = 1$$

25. The solutions are $s = 15$ and $s = -45$.

$$s + 15 = 30 \qquad\qquad s + 15 = -30$$
$$s = 30 - 15 \qquad\qquad s = -30 - 15$$
$$s = 15 \qquad\qquad s = -45$$

26. The solutions are $m = 5$ and $m = -\dfrac{7}{3}$.

$$3m - 4 = 11 \qquad\qquad 3m - 4 = -11$$
$$3m = 15 \qquad\qquad 3m = -7$$
$$m = 5 \qquad\qquad m = -\frac{7}{3}$$

27. The solutions are $t = 28$ and $t = -28$.

$$t \div 7 = 4 \qquad\qquad t \div 7 = -4$$
$$7 \times \left(\frac{t}{7}\right) = 4 \times 7 \qquad\qquad 7 \times \left(\frac{t}{7}\right) = -4$$
$$t = 4 \times 7 \qquad\qquad t = -4 \times 7$$
$$t = 28 \qquad\qquad t = -28$$

28. The solutions are $a = 3$ and $a = -\dfrac{16}{3}$.

$$6a + 7 = 25 \qquad\qquad 6a + 7 = -25$$
$$6a = 18 \qquad\qquad 6a = -32$$
$$a = 3 \qquad\qquad a = \frac{-32}{6}$$
$$a = \frac{-16}{3}$$

29. The solutions are $g = \dfrac{9}{5}$ and $g = -\dfrac{9}{5}$.

$$4g \times 5 = 36 \qquad\qquad 4g \times 5 = -36$$

$$\dfrac{4g \times 5}{5} = \dfrac{36}{5} \qquad\qquad \dfrac{4g \times 5}{5} = \dfrac{-36}{5}$$

$$4g = \dfrac{36}{5} \qquad\qquad 4g = \dfrac{-36}{5}$$

$$\left(\dfrac{1}{4}\right) \times (4g) = \left(\dfrac{36}{5}\right) \times \left(\dfrac{1}{4}\right) \qquad \left(\dfrac{1}{4}\right) \times (4g) = \left(\dfrac{-36}{5}\right) \times \left(\dfrac{1}{4}\right)$$

$$g = \dfrac{9}{5} \qquad\qquad g = -\dfrac{9}{5}$$

30. The solutions are $b = 2$ and $b = -2$.

$$14b \div 7 = 4 \qquad\qquad 14b \div 7 = -4$$

$$\dfrac{14b}{7} = 4 \qquad\qquad \dfrac{14b}{7} = -4$$

$$7 \times \left(\dfrac{14b}{7}\right) = 4 \times 7 \qquad 7 \times \left(\dfrac{14b}{7}\right) = -4 \times 7$$

$$14b = 28 \qquad\qquad 14b = -28$$

$$b = 2 \qquad\qquad b = -2$$

Interval notation

31. The correct answer is $3 \le x \le 16$.

32. The correct answer is $-5 < x < 2$.

33. The correct answer is $-11 \le x < -7$.

34. The correct answer is $a < x \le b$.

35. The correct answer is $2s \le x \le s^2 - 3$.

36. The correct answer is $(0, 12)$.

37. The correct answer is $[-9, 9]$.

38. The correct answer is $\left(\dfrac{1}{2}, 21\right]$.

39. The correct answer is $[a, 7b]$.

40. The correct answer is $\left(\dfrac{a+2}{b}, \dfrac{a}{c}\right]$.

Chapter Review Solutions

1. The correct answer is $24a^6 + 18a^3$.

$$6a^3\left(4a^3 + 3\right) = \left(6a^3 \times 4a^3\right) + \left(6a^3 \times 3\right)$$
$$= \left(24a^{3+3}\right) + \left(18a^3\right)$$
$$= 24a^6 + 18a^3$$

2.

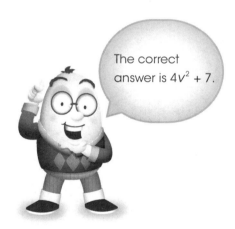

The correct answer is $4v^2 + 7$.

$$\frac{28v^2 + 49}{7} = \frac{28v^2}{7} + \frac{49}{7}$$
$$= 4v^2 + \frac{49}{7}$$
$$= 4v^2 + 7$$

3. The correct answer is $c^2 + 16c + 60$.

$$(c + 10)(c + 6) = (c \times c) + (c \times 6) + (10 \times c) + (10 \times 6)$$
$$= c^2 + 6c + 10c + 60$$
$$= c^2 + 16c + 60$$

4. The correct answer is $36s^2 + 26s - 42$.

$$(4s + 6)(9s - 7) = (4s \times 9s) + \left[4s \times (-7)\right] + (6 \times 9s) + \left[6 \times (-7)\right]$$
$$= \left(36s^2\right) + (-28s) + (54s) + (-42)$$
$$= 36s^2 - 28s + 54s - 42$$
$$= 36s^2 + 26s - 42$$

5. The correct answer is $42x^5 - 21x^4 + 24x^3 + 2x^2 + 8$.

$$\left(7x^2 + 4\right)\left(6x^3 - 3x^2 + 2\right) = \left(7x^2 \times 6x^3\right) + \left\lfloor 7x^2 \times \left(-3x^2\right)\right\rfloor + \left(7x^2 \times 2\right) + \left(4 \times 6x^3\right) + \left\lfloor 4 \times \left(-3x^2\right)\right\rfloor + \left(4 \times 2\right)$$
$$= \left(42x^{2+3}\right) + \left(-21x^{2+2}\right) + \left(14x^2\right) + \left(24x^3\right) + \left(-12x^2\right) + \left(8\right)$$
$$= 42x^5 - 21x^4 + 14x^2 + 24x^3 - 12x^2 + 8$$
$$= 42x^5 - 21x^4 + 24x^3 + 14x^2 - 12x^2 + 8$$
$$= 42x^5 - 21x^4 + 24x^3 + 2x^2 + 8$$

6. The factors of the binomial are t and $(3t + 7)$.

7. The factors of the binomial are h and $(-h + 5)$.

8. The factors of the trinomial are $(k + 2)$ and $(k - 12)$.

9. The factors of the trinomial are $(-a + 3)$ and $(a - 3)$.

10. The factors of the trinomial are $(z - 10)$ and $(2z + 7)$.

11. The solutions are $a = 5$ and $a = 3$.

$$\begin{array}{ll}
3a - 12 = 3 & \qquad 3a - 12 = -3 \\
3a = 15 & \qquad 3a = 9 \\
a = 5 & \qquad a = 3
\end{array}$$

12. The solutions are $z = \dfrac{8}{3}$ and $z = -\dfrac{8}{3}$.

$$\begin{array}{ll}
2z \times 9 = 48 & \qquad 2z \times 9 = -48 \\
18z = 48 & \qquad 18z = -48 \\
z = \dfrac{48}{18} & \qquad z = \dfrac{-48}{18} \\
z = \dfrac{8}{3} & \qquad z = -\dfrac{8}{3}
\end{array}$$

13. The solutions are $d = 16$ and $d = -16$.

$$\begin{array}{ll}
4d \div 16 = 4 & \qquad 4d \div 16 = -4 \\
\dfrac{4d}{16} = 4 & \qquad \dfrac{4d}{16} = -4 \\
16 \times \left(\dfrac{4d}{16}\right) = 4 \times 16 & \qquad 16 \times \left(\dfrac{4d}{16}\right) = -4 \times 16 \\
4d = 64 & \qquad 4d = -64 \\
d = 16 & \qquad d = -16
\end{array}$$

14.

The solutions are $b = 2$ and $b = -2$.

$$7b \times 3 = 42 \qquad\qquad 7b \times 3 = -42$$
$$21b = 42 \qquad\qquad 21b = -42$$
$$b = 2 \qquad\qquad b = -2$$

15. The solutions are $j = \dfrac{13}{3}$ and $j = -\dfrac{22}{3}$.

$$6j + 9 = 35 \qquad\qquad 6j + 9 = -35$$
$$6j = 26 \qquad\qquad 6j = -44$$
$$j = \dfrac{26}{6} \qquad\qquad j = \dfrac{-44}{6}$$
$$j = \dfrac{13}{3} \qquad\qquad j = -\dfrac{22}{3}$$

16. The correct answer is $-2 < x < 7$.

17. The correct answer is $12 \le x \le 23$.

18. The correct answer is $-4 < x \le 2$.

19. The correct answer is $y \le x \le 3z$.

20. The correct answer is $6 + t < x \le 2t^2 + 7$.

21. The correct answer is $[-1, 7)$.

22. The correct answer is $(6, 14]$.

23. The correct answer is $\left(\dfrac{3}{7}, 9\right)$.

24. The correct answer is $[2c, 3d)$.

25. The correct answer is $\left[2(a - 4), \dfrac{a + b}{3}\right]$.

Chapter 2

Graphing

In this chapter, we'll review the following concepts:

The coordinate plane
Graphing linear equations
Finding the slope of a line
The slope-intercept formula
Graphing quadratic equations

The coordinate plane

A coordinate plane is a grid upon which graphs are drawn. A typical coordinate plane looks like this:

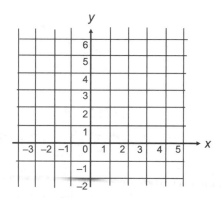

The numbers on the grid show the units by which the coordinate plane is measured. They extend from zero to infinity in both positive and negative directions.

The coordinate plane consists of two axes, the *x*-axis and the *y*-axis. The *x*-axis extends horizontally across the plane, as shown by the letter *x* in the figure, and the *y*-axis extends vertically, as shown by the letter *y*. The point located at the intersection of the *x*- and *y*-axes is known as the **origin.**

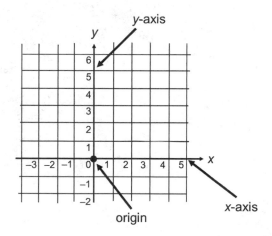

To show location on the coordinate plane, we draw in points known as **coordinates.** The location of a point is determined by the *x*- and *y*-values that make up the coordinate. A pair of *x*- and *y*-coordinates is known as an **ordered pair** and is written in parentheses as follows:

(*x*-value, *y*-value)

An ordered pair such as (2, 3) indicates a point on the coordinate plane located at 2 on the x axis and 3 on the y-axis. To plot the point, count 2 units to the right from the origin and 3 units up:

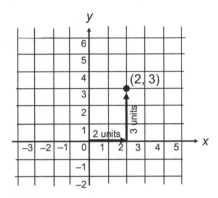

To identify the location of any point on the coordinate plane, we determine its x- and y-coordinates to create an ordered pair

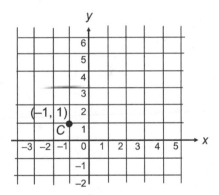

In the figure shown, point C is located at coordinates (–1, 1).

The origin is
located at (0, 0).

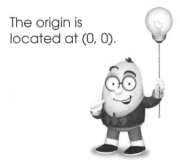

Practice Questions—The coordinate plane

Directions: Plot the points shown in the coordinate plane. You will find the Practice Question Solutions on page 39.

1. *A* (3, 2) and *B* (2, 3)

2. *Y* (1, 4) and *Z* (4, 1)

3.

P (2, –1) and *Q* (–1, 2)

4. *C* (0, 5) and *D* (5, 0)

5. *F* (–1, –2) and G (–3, –1)

Directions: Write the coordinates for the points shown in the graphs below. You will find the Practice Question Solutions on page 39.

6.

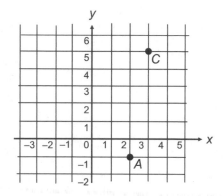

7.

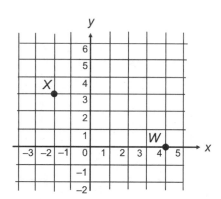

8.

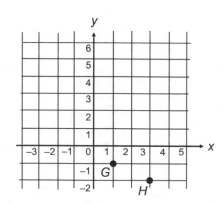

egghead's Guide to Calculus

9.   **10.**

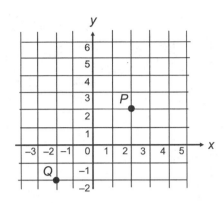

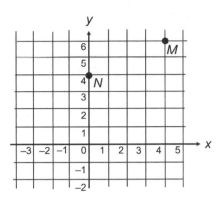

Graphing linear equations

Any two points in the coordinate plane can be connected to form a line. To graph a line in the coordinate plane, we simply connect two points:

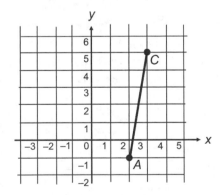

The line can stop at the two endpoints, as shown above, or it can go on forever:

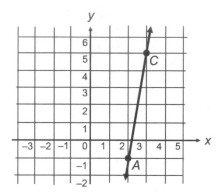

The point where a line crosses the *x*-axis in the graph is known as the **x-intercept.** The point where the line crosses the *y*-axis is the **y-intercept.** In the figure shown below, the *x*-intercept is (1, 0). The *y*-intercept is (0, 2).

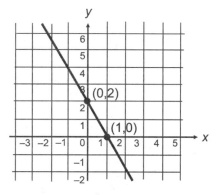

A **linear equation** is an equation whose graph can be represented in the coordinate plane as a straight line. The equation $y = 2x + 1$, for instance, is a linear equation. There are several ways to determine the graph of a linear equation. One way is by developing a table of values and then plotting the points in the table.

To generate a **table of values,** determine the value of *y* for two or more values of *x*. With the equation $y = 2x + 1$, we could substitute several values for *x* and get the following values for *y*:

x	y
-2	-3
-1	-1
0	1
1	3
2	5

The graph of the equation would look like this:

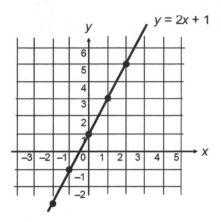

The line crosses the y-axis at coordinate (0, 1), so this point is the y-intercept.

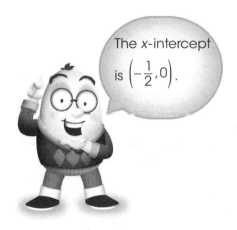

Practice Questions—Graphing linear equations

Directions: You will find the Practice Question Solutions to the questions below on page 40.

11. What is the *y*-intercept of the line shown?

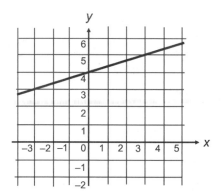

12. What is the *x*-intercept of the line shown?

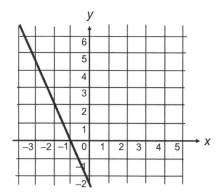

13. Graph the equation of the line $y = x + 3$.

14. Graph the equation of the line $y = 2x - 2$.

15. Graph the equation of the line $y = -4x + 3$.

Finding the slope of a line

In the coordinate plane, the **slope** of a line is a measure of how steep the line is. The measure of slope is symbolized by the letter m. The slope of a line can be determined using the following formula:

$$m = \frac{\Delta y}{\Delta x}$$

In this formula, the Greek letter Δ (delta) is used to represent the word "change." So, the slope is a measure of the change in y divided by the change in x.

We can find the slope of a line using the x- and y-coordinates of any two points. Take the points $(1, 2)$ and $(2, 3)$, for instance. Using the formula for the slope of a line, we would take the difference in the y-values and divide it by the difference in the x-values:

$$m = \frac{\Delta y}{\Delta x}$$
$$= \frac{(3 - 2)}{(2 - 1)}$$
$$= \frac{1}{1}$$

The slope of the line is 1.

In the numerator of the fraction, we took the y-value of the second point, 3, and subtracted the y-value of the first point, 2. The difference is $3 - 2 = 1$. In the denominator, we took the x-value of the second point, 2, and subtracted the x-value of the first point, 1. The difference is $2 - 1 = 1$.

Another way to write the slope formula is by calling the two points (x_1, y_1) and (x_2, y_2). We then plug these values into the slope formula as follows:

$$m = \frac{y_2 - y_1}{x_2 - x_1}$$

The value $y_2 - y_1$ represents the change in y, and the value $x_2 - x_1$ represents the change in x.

Practice Questions—Finding the slope of the line

Directions: Find the slope of the line that connects each pair of points below. You will find the Practice Question Solutions on page 40.

16. A (6, 4) and B (3, 3)

17. S (1, 10) and T (4, –5)

18. L (20, –7) and M (12, 1)

Directions: Find the slope of the line in each graph shown. You will find the Practice Question Solutions on page 41.

19.

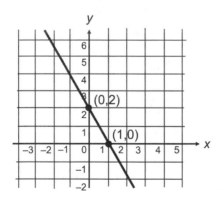

20.

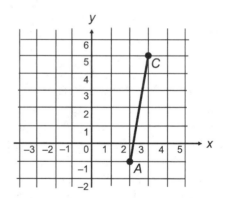

The slope-intercept formula

The equation of a line is often written in the following form:

$$y = mx + b$$

In this standard format, the letter m represents the slope of the line. The letter b represents the y-intercept.

In the equation $y = 4x + 2$, for example, the number 4 represents the slope, m, and the number 2 represents the y-intercept, b. The y-intercept tells us that the line crosses the y-axis at the point $(0, 2)$.

The y-intercept can be written as $(0, b)$.

This standard format for the equation of a line is known as the **slope-intercept formula**. We can use the slope-intercept formula to determine the equation of a line given only its slope and its y-intercept. Or, we can use the formula to find the equation of a line given its slope and one point on the line.

Practice Questions—The slope-intercept formula

Directions: Find the slope-intercept formula for each question below. You will find the Practice Question Solutions on page 41.

21. What is the equation of a line with slope 3 and a *y*-intercept of 7?

22. What is the equation of a line with slope –4 and a *y*-intercept of 2?

23. What is the equation of a line with slope $\frac{1}{2}$ and a *y*-intercept of –5?

24. What is the equation of a line with slope 1 that passes through the point (2, 4)?

25. What is the equation of a line with slope –2 that passes through the point (–1, 3)?

Graphing quadratic equations

Quadratic equations are those that contain a variable raised to the second power. In their standard form, they are written as follows:

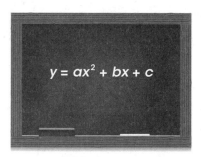

$$y = ax^2 + bx + c$$

The equation $y = x^2$ is the simplest form of a quadratic equation. Other examples are listed below:

$$y = 2x^2 + 3x + 9$$
$$y = 7x^2 - 4x$$
$$y = -x^2 + 2x - 5$$

To graph a quadratic equation, we can create a table of values and plot the points found. It's a good idea to plot 5 or more points, so we can see the shape of the graph accurately.

Using the equation $y = x^2 + 2x - 1$, plug in at least 5 values for x and solve for y. The results are as follows:

x	y
−4	7
−3	2
−2	−1
−1	−2
0	−1
1	2
2	7
3	14

The graph appears as a curve with its lowest point at the coordinate (−1, 2). We call this low point the **vertex** of the curve.

The curve itself is called a **parabola**

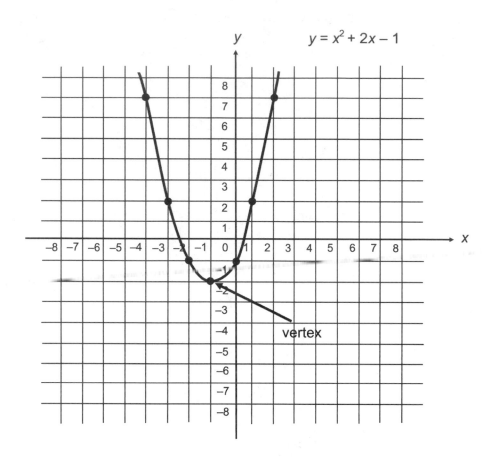

$$y = x^2 + 2x - 1$$

vertex

Practice Questions—Graphing quadratic equations

Directions: Draw the graphs of the quadratic equations below. You will find the Practice Question Solutions on page 42.

26. $y = x^2$

27. $y = -x^2$

28. $y = x^2 - 1$

29. $y = x^2 - 2x + 1$

30. $y = x^2 + 4x - 1$

Chapter Review

Directions: Plot the points shown in the coordinate plane. You will find the solutions on page 45.

1. J (1, 6) and K (–1, 3)

2. R (5, 4) and S (2, –2)

3.

X (4, 3) and Y (1, 2)

4. E (–2, 3) and F (5, 5)

5. L (–1, 2) and M (4, –1)

6. What is the y-intercept of the line shown?

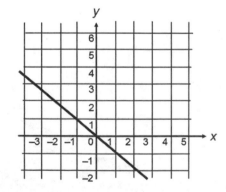

7. Graph the equation of the line $y = 2x + 2$.

8. Graph the equation of the line $y = 3x – 4$.

9. Find the slope of the line that connects the pair of points P (–2, 16) and Q (4, –8).

Using the slope formula, find the slope of the line in each graph shown.

10.

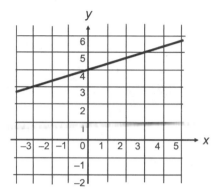

11.

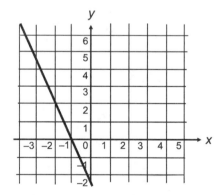

12. What is the equation of a line with slope $\frac{2}{3}$ and a y-intercept of 6?

13. What is the equation of a line with slope –3 that passes through the point (1, –5)?

Draw the graphs of the quadratic equations shown.

14. $y = 2x^2$

15. $y = \frac{1}{2}x^2 - 1$

Practice Question Solutions

The coordinate plane

1. The correct answer is shown below.

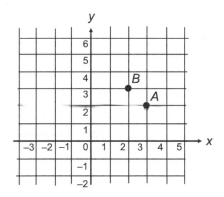

2. The correct answer is shown below.

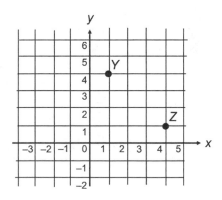

3. The correct answer is shown below

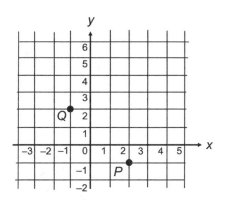

4. The correct answer is shown below.

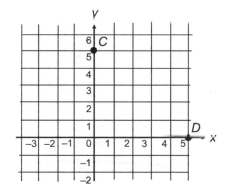

5. The correct answer is shown below.

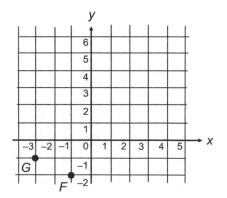

6. The correct answers are *A* (2, –1) and *C* (3, 5).

7. The correct answers are *W* (4, 0) and *X* (–2, 3).

8. The correct answers are *G* (1, –1) and *H* (3, –2).

9. The correct answers are *P* (2, 2) and *Q* (–2, –2).

10. The correct answers are *M* (4, 6) and *N* (0, 4).

Graphing linear equations

11. The *y*-intercept of the line is (0, 4).

12. The *x*-intercept of the line is (–1, 0).

13. The correct answer is shown below.

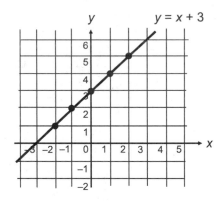

$y = x + 3$

14.

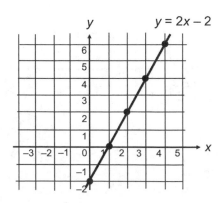

$y = 2x - 2$

15. The correct answer is shown below.

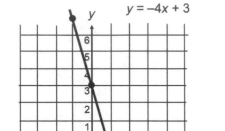

$y = -4x + 3$

Finding the slope of the line

16. The correct answer is $\frac{1}{3}$.

$$m = \frac{y_2 - y_1}{x_2 - x_1}$$
$$= \frac{(3) - (4)}{(3) - (6)}$$
$$= \frac{-1}{-3}$$
$$= \frac{1}{3}$$

17. The correct answer is –5.

$$m = \frac{y_2 - y_1}{x_2 - x_1}$$
$$= \frac{(-5) - (10)}{(4) - (1)}$$
$$= \frac{-15}{3}$$
$$= -5$$

18. The correct answer is –1.

$$m = \frac{y_2 - y_1}{x_2 - x_1}$$
$$= \frac{(1) - (-7)}{(12) - (20)}$$
$$= \frac{8}{-8}$$
$$= -1$$

19. The correct answer is –2.

The line passes through the points (0, 2) and (1, 0). Substitute these points into the slope formula:

$$m = \frac{y_2 - y_1}{x_2 - x_1}$$
$$= \frac{(0) - (2)}{(1) - (0)}$$
$$= \frac{-2}{1}$$
$$= -2$$

20. The correct answer is 6.

The line passes through the points (2, –1) and (3, 5). Substitute these points into the slope formula:

$$m = \frac{y_2 - y_1}{x_2 - x_1}$$
$$= \frac{(5) - (-1)}{(3) - (2)}$$
$$= \frac{6}{1}$$
$$= 6$$

The slope-intercept formula

21. The correct answer is $y = 3x + 7$.

22. The correct answer is $y = -4x + 2$.

23. The correct answer is $\frac{1}{2}x - 5$.

24. The correct answer is $y = x + 2$.

Plug the given values into the slope-intercept equation and solve for b:

$$y = mx + b$$
$$(4) = (1)(2) + b$$
$$4 = 2 + b$$
$$2 = b$$

The value of $m = 1$, and the value of $b = 2$.

25. The correct answer is $y = -2x + 1$.

Plug the given values into the slope-intercept equation and solve for b:

$$y = mx + b$$
$$(3) = (-2)(-1) + b$$
$$3 = 2 + b$$
$$1 = b$$

The value of $m = -2$, and the value of $b = 1$.

Graphing quadratic equations

26. The correct answer is shown below.

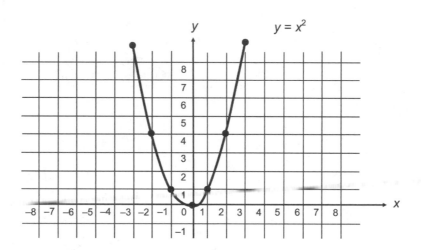

27. The correct answer is shown below.

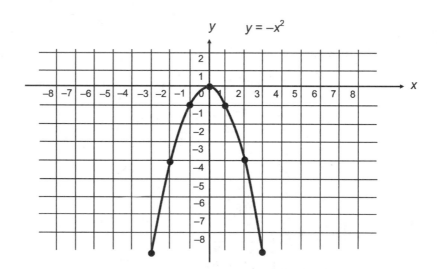

28. The correct answer is shown below.

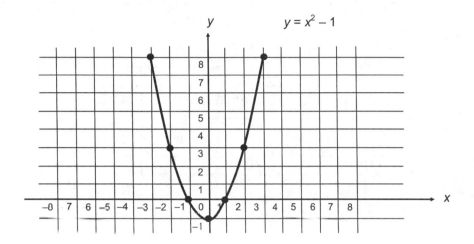

$y = x^2 - 1$

29. The correct answer is shown below.

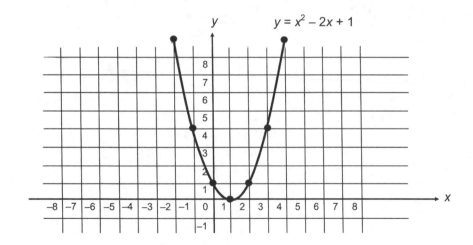

$y = x^2 - 2x + 1$

30. The correct answer is shown below.

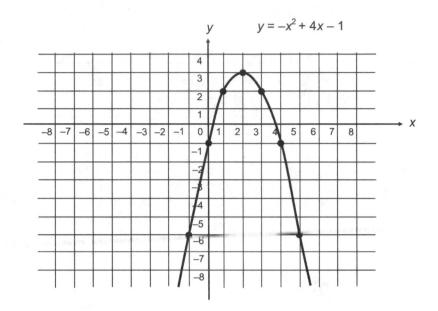

$$y = -x^2 + 4x - 1$$

Chapter Review Solutions

1. The correct answer is shown below.

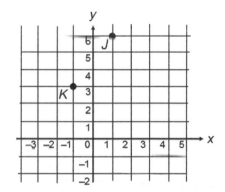

2. The correct answer is shown below.

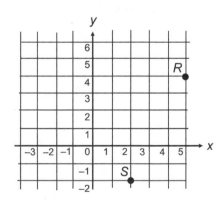

3. The correct answer is shown below.

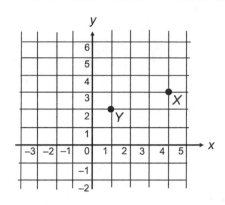

4. The correct answer is shown below.

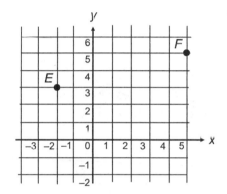

5. The correct answer is shown below.

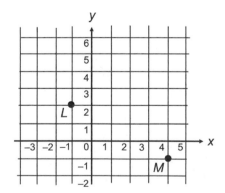

6. The *y*-intercept of the line is (0, 0).

7. The correct answer is shown below.

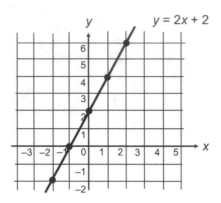

8. The correct answer is shown below.

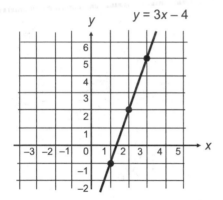

9. The correct answer is –4.

$$m = \frac{y_2 - y_1}{x_2 - x_1}$$

$$= \frac{(-8) - (16)}{(4) - (-2)}$$

$$= \frac{-24}{6}$$

$$= -4$$

10. The correct answer is $\frac{1}{3}$.

The line passes through the points (–3, 3) and (0, 4). Substitute these points into the slope formula:

$$m = \frac{y_2 - y_1}{x_2 - x_1}$$

$$= \frac{(4) - (3)}{(0) - (-3)}$$

$$= \frac{1}{3}$$

11. The correct answer is –2.

The line passes through the points (–1, 0) and (–2, 2). Substitute these points into the slope formula:

$$m = \frac{y_2 - y_1}{x_2 - x_1}$$

$$= \frac{(2) - (0)}{(-2) - (-1)}$$

$$= \frac{2}{-1}$$

$$= -2$$

Any two points on the line could be used in the formula.

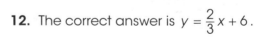

12. The correct answer is $y = \frac{2}{3}x + 6$.

13. The correct answer is $y = -3x - 2$.

Plug the given values into the slope-intercept equation and solve for b:

$$y = mx + b$$
$$(-5) = (-3)(1) + b$$
$$-5 = -3 + b$$
$$-2 = b$$

The value of $m = -3$, and the value of $b = -2$.

14.

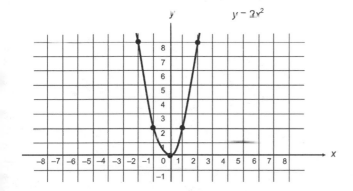

15. The correct answer is shown below

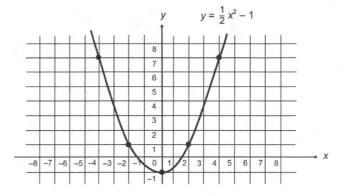

Chapter 3

Functions

In this chapter, we'll review the following concepts:

What is a function?
Domain and range
Functional notation
Composite functions
Piecewise functions

What is a function?

Along with graphing, another concept used throughout calculus is that of functions. In mathematics, a **function** is a specific type of relationship between two values. One of the values serves as an input, and the other value is the output. An operation is performed on the first value, the input, to produce the second value, the output.

Domain and range

In mathematics, the **domain** of a function includes all of the input values for that function. The **range** of the function includes all of the output values for the function.

In order for a relationship between values to be classified as a function, each input can have only one output. In other words, for each value in the domain, there can be only one corresponding value in the range.

Here is an example:

x	y
–2	–2
–1	–1
0	0
1	2
2	3

In the table shown, the x-values represent the domain. The y-values represent the range. For each x-value in the table, there is only one corresponding y-value.

In the table below, however, some of the x-values have more than one y-value.

x	y
–2	–2
–2	2
–1	–1
–1	1
0	0

The relationship is therefore a function.

The x-value –2 has two corresponding y-values: –2 and 2. The x-value 1 also has two corresponding y-values, –1 and 1. This relationship is not a function.

It is acceptable for the range of a function to have more than one corresponding value in the domain of the function. In other words, for each value of y, there can be more than one value of x, as shown in the table:

x	y
–2	2
–1	2
0	2
1	2
2	2

In this table, the y-value of 2 has 5 different corresponding values of x: 2, 1, 0, 1, and 2. Each x-value, however, only has one corresponding y-value, so the relationship is still a function.

Graphing functions

When we draw the graph of a function, the domain is represented on the x-axis.

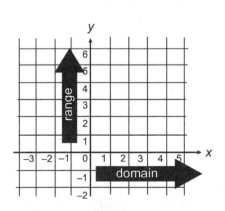

The range is represented on the y-axis.

As an example, here is a graph of the function represented by the coordinates shown in the table below:

x	y
−2	−2
−1	−1
0	0
1	2
2	3

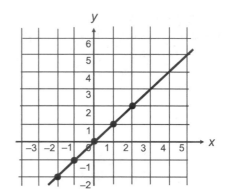

The domain values are shown on the x-axis, and the range values are shown on the y-axis.

The vertical line test

In order to test whether a graph represents a function, we can use the **vertical line test.** If a vertical line can be drawn that passes through the graph at more than one point, then the graph does not represent a function.

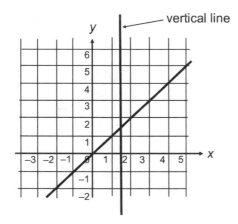

This graph does represent a function.

A vertical line drawn through the graph above can only ever intersect the graphed line at a single point.

This graph of a circle, on the other hand, does **not** represent a function:

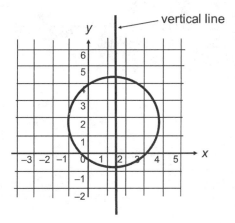

Using the vertical line test, we see that it's possible for the vertical line to intersect the circle at more than one point.

When in doubt, use the vertical line test!

Practice Questions—Domain and range

Directions: Determine whether each graph or set of values below represents a function. You will find the Practice Question Solutions on page 64.

1.

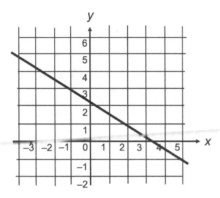

2.

x	y
−3	7
−2	5
−1	3
0	1
1	−1

3.

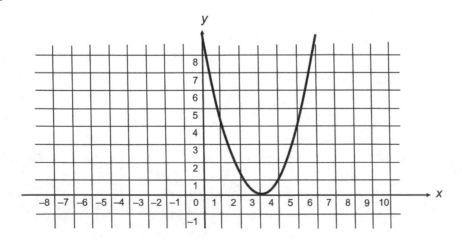

4.

x	y
10	5
20	10
30	15
30	20
40	25
50	30
60	35

5.

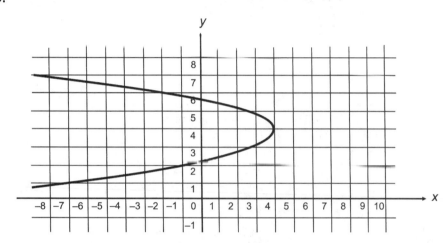

Functional notation

To name a function, we use the following notation:

The letter *f* is the name of the function. The value in parentheses describes the input value, or **independent variable,** of the function.

The input value is also called the **argument** of the function.

Functions are often written as equations. The value on the right side of the equals sign describes the output value, or **dependent variable,** of the function.

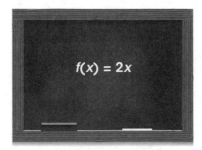

In this case, the input value of the function is *x*, and the output value is 2*x*. For every input value of *x* that you plug into the function, you get a corresponding output value of 2*x*.

Example

If $f(x) = 2x$, find the value of $f(x)$ when x equals 4.

To solve this problem, we plug 4 in for x on the right side of the equation. When $x = 4$, the value of $f(x)$ is 2×4, or 8.

The phrase $f(x)$ is pronounced "f of x."

Using other variables

The phrase $f(x)$ is a common way to name a function. The letters f and x can be replaced by other letters, however, such as $g(a)$ or $h(c)$. The function still means the same thing.

The function $g(a) = 2a$ describes the same operation as $f(x) = 2x$. In both cases, the function involves taking the input value and multiplying it by 2.

Practice Questions—Functional notation

Directions: Set up and solve the following functions. You will find the Practice Question Solutions on page 64.

6. If $f(x) = 3x$, find the value of $f(x)$ when x equals 9.

7. Evaluate the function $f(x) = x + 1$ at $x = 5$.

8. If $f(x) = x - 7$, find the value of $f(x)$ when x equals 12.

9. If $g(a) = a^2$, what is the value of $g(a)$ when a equals 3?

10. Evaluate the function $h(c) = c^2 + 3$ at $c = 2$.

Composite functions

Some functions are written as **composite functions.** These are functions that join together one or more functions.

Composite functions are functions within functions.

Examples

Consider the following composite function problem:

If $f(x) = x^2$ and $g(x) = 2x$, find $g(f(x))$.

To solve this problem, we first set up the function $g(x) = 2x$ using parentheses:

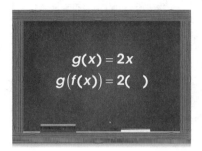

Substitute the value of $f(x)$ in the parentheses. We are told that $f(x) = x^2$, so we insert x^2 in the parentheses:

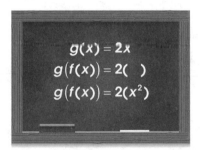

The value of $g(f(x))$ is $2x^2$.

As another example, let's solve for $f(g(x))$. In this composite function, $f(x) = x^2$ and $g(x) = 2x$. This time, we insert $2x$ in the parentheses:

$$f(x) = x^2$$
$$f(g(x)) = (\quad)^2$$
$$f(g(x)) = (2x)^2$$
$$f(g(x)) = 4x^2$$

The composite function $f(g(x))$ can also be written as $(f \circ g)(x)$.

Practice Questions—Composite functions

Directions: Solve the following functions for the indicated values. You will find the Practice Question Solutions on page 65.

11. If $f(x) = x - 1$ and $g(x) = x^2$, find $f(g(x))$.

12. If $f(x) = x + 2$ and $g(x) = x - 7$, find $g(f(x))$.

13. If $f(x) = 3x$ and $g(x) = 2x$, find $f(g(x))$.

14. If $g(x) = 2x^2$ and $h(x) = x + 1$, find $h(g(x))$.

15. If $f(a) = 7a$ and $j(a) = 2a$, find $f(j(a))$.

Piecewise functions

A **piecewise function** is a function whose graph is made up of pieces. Here is an example of a graph of a piecewise function:

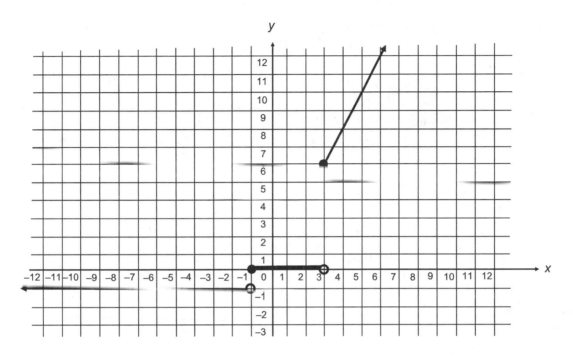

To describe this function using interval notation, we would write the following:

$$f(x) = \begin{cases} -1, & (-\infty, -1) \\ 0, & (-1, 3) \\ 2x, & (3, \infty) \end{cases}$$

Both descriptions mean the same thing.

We could also describe the function using inequalities:

$$f(x) = \begin{cases} -1, & x < -1 \\ 0, & -1 \le x < 3 \\ 2x, & x \ge 3 \end{cases}$$

Practice Questions—Piecewise functions

Directions: Evaluate the following functions and answer each question. You will find the Practice Question Solutions on page 65.

16. Draw a graph of the following function:

$$f(x) = \begin{cases} -1, & (-\infty, 1) \\ x, & (1, \infty) \end{cases}$$

17. Describe the following function using inequalities:

$$f(x) = \begin{cases} -4, & (-\infty, -4) \\ 0, & (-4, 1) \\ x + 1, & (1, \infty) \end{cases}$$

18. Given the function $g(x)$ shown below, find the value of $g(x)$ when x equals 3.

$$g(x) = \begin{cases} x + 1, & x < 0 \\ x + 2, & x \geq 0 \end{cases}$$

19. Evaluate the function $h(a)$ shown below at $a = 1$.

$$h(a) = \begin{cases} -3, & (-\infty, 2) \\ a - 3, & (2, \infty) \end{cases}$$

20. Given the function $j(x)$ shown below, what is the value of $j(x)$ when x equals -4?

$$j(x) = \begin{cases} x, & (-\infty, 1) \\ 4x, & (1, \infty) \end{cases}$$

Chapter Review

Solutions can be found on page 66.

1. Does the graph below represent a function?

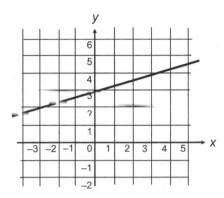

2. Determine whether the table of values below represents a function.

x	y
5	25
4	16
3	9
2	4
1	2
1	1
0	0

3. If $f(x) = 2x + 3x$, find the value of $f(x)$ when x equals 1.

4. If $g(x) = 7x^2$, what is the value of $g(x)$ when x equals 10?

5. Evaluate the function $h(a) = 3a(a - 2)$ at $a = 4$.

6. If $f(x) = x + 3$ and $g(x) = 2x$, find $g(f(x))$.

7. If $f(x) = 5x$ and $g(x) = 3x$, find $f(g(x))$.

8. If $f(x) = x^2$ and $g(x) = 2x^2$, find $g(f(x))$.

9. If $g(c) = 3c$ and $h(c) = c + 1$, find $g(h(c))$
.

10. If $f(n) = 12n$ and $g(n) = 2n$, find $g(f(n))$.

11. Draw a graph of the following function:

$$h(a) = \begin{cases} 0, & (-\infty, 0) \\ x^2, & (0, \infty) \end{cases}$$

12. Describe the following function using interval notation:

$$g(x) = \begin{cases} 3, & x \leq 1 \\ 4, & 1 < x \leq 3 \\ 5, & x > 3 \end{cases}$$

13. Given the function $f(x)$ shown below, find the value of $f(x)$ when x equals 4.

$$f(x) = \begin{cases} -x, & x < 3 \\ 0, & x = 3 \\ x, & x > 3 \end{cases}$$

14. Evaluate the function $g(c)$ shown below at $c = -3$.

$$g(c) = \begin{cases} c + 1, & (-\infty, 0) \\ c - 1, & (0, \infty) \end{cases}$$

15. Given the function $h(x)$ shown below, what is the value of $h(x)$ when x equals 5?

$$h(x) = \begin{cases} x, & (-\infty, 4) \\ x^2, & (4, 5) \\ x^3, & (5, \infty) \end{cases}$$

Practice Question Solutions

Domain and range

1. The graph shown represents a function

A vertical line drawn through the graph could only intersect the line at a single point. The graph passes the vertical line test, so it is a function.

2. The table of values represents a function.

For each *x*-value in the table, there is only one corresponding *y*-value, so the relationship between the values in the table is a function.

3. The graph shown represents a function.

A vertical line drawn through the graph could only intersect the parabola at a single point.

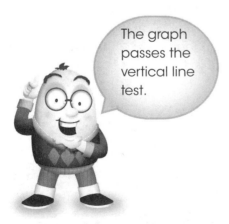

The graph passes the vertical line test.

4. The table of values does not represent a function.

For one of the *x*-values in the table, there more than one corresponding *y*-value. The *x*-value 30 has two corresponding *y*-values: 15 and 20. Therefore, the table of values does not reflect a function.

5. The graph shown does not represent a function.

A vertical line drawn through the graph could possibly intersect the parabola at two points. The graph does not pass the vertical line test, so it is not a function.

Functional notation

6. The correct answer is 27.

$f(x) = 3x$
$f(9) = 3(9)$
$\quad = 27$

7. The correct answer is 6.

$f(x) = x + 1$
$f(5) = (5) + 1$
$\quad = 6$

8. The correct answer is 5.

$f(x) = x - 7$
$f(12) = (12) - 7$
$\quad = 5$

9. The correct answer is 9.

$g(a) = a^2$
$g(3) = (3)^2$
$\quad = 9$

10. The correct answer is 7.

$h(c) = c^2 + 3$
$h(2) = (2)^2 + 3$
$\quad = 4 + 3$
$\quad = 7$

Composite functions

11. The correct answer is $x^2 - 1$.

$$f(x) = x - 1$$
$$f(g(x)) = (\quad) - 1$$
$$f(g(x)) = (x^2) - 1$$
$$f(g(x)) = x^2 - 1$$

12. The correct answer is $x - 5$.

$$g(x) = x - 7$$
$$g(f(x)) = (\quad) - 7$$
$$g(f(x)) = (x + 2) - 7$$
$$g(f(x)) = x - 5$$

13. The correct answer is $6x$.

$$f(x) = 3x$$
$$f(g(x)) = 3(\quad)$$
$$f(g(x)) = 3(2x)$$
$$f(g(x)) = 6x$$

14.

The correct answer is $2x^2 + 1$.

$$h(x) = x + 1$$
$$h(g(x)) = (\quad) + 1$$
$$h(g(x)) = (2x^2) + 1$$
$$h(g(x)) = 2x^2 + 1$$

15. The correct answer is $14a$.

$$f(a) = 7a$$
$$f(j(a)) = 7(\quad)$$
$$f(j(a)) = 7(2a)$$
$$f(j(a)) = 14a$$

Piecewise functions

16. The correct answer is shown below.

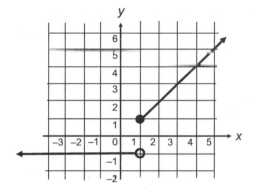

17. The correct answer is shown below.

$$f(x) = \begin{cases} -4, & x < -4 \\ 0, & -4 \le x < 1 \\ x + 1, & x \ge 1 \end{cases}$$

18. The correct answer is 5.

$$g(3) = (3) + 2$$
$$g(3) = 5$$

19. $h(1) = -3$

20. The correct answer is -4.

$$j(-4) = (-4)$$

Chapter Review Solutions

1. The graph shown represents a function.

   A vertical line drawn through the graph could only intersect the line at a single point. The graph passes the vertical line test, so it is a function.

2. The table of values does not represent a function.

   For one of the *x*-values in the table, there is more than one corresponding *y*-value. The *x*-value 1 has two corresponding *y*-values: 2 and 1. Therefore, the table of values does not reflect a function.

   Each *x*-value in a function can have only one *y*-value.

3. The correct answer is 5.

   $$f(x) = 2x + 3x$$
   $$f(1) = 2(1) + 3(1)$$
   $$= 2 + 3$$
   $$= 5$$

4. The correct answer is 700.

   $$g(x) = 7x^2$$
   $$g(10) = 7(10)^2$$
   $$= 7(100)$$
   $$= 700$$

5. The correct answer is 24.

   $$h(a) = 3a(a - 2)$$
   $$h(4) = 3(4)[(4) - 2]$$
   $$= 12(2)$$
   $$= 24$$

6. The correct answer is $2x + 6$.

   $$g(x) = 2x$$
   $$g(f(x)) = 2(\quad)$$
   $$g(f(x)) = 2(x + 3)$$
   $$g(f(x)) = 2x + 6$$

7. The correct answer is $15x$.

   $$f(x) = 5x$$
   $$f(g(x)) = 5(\quad)$$
   $$f(g(x)) = 5(3x)$$
   $$f(g(x)) = 15x$$

8. The correct answer is $2x^4$.

   $$g(x) = 2x^2$$
   $$g(f(x)) = 2(\quad)^2$$
   $$g(f(x)) = 2(x^2)^2$$
   $$g(f(x)) = 2x^4$$

9. The correct answer is $3c + 3$.

   $$g(c) = 3c$$
   $$g(h(c)) = 3(\quad)$$
   $$g(h(c)) = 3(c + 1)$$
   $$g(h(c)) = 3c + 3$$

10. The correct answer is 24n.

$$g(n) = 2n$$
$$g(f(n)) = 2(\quad)$$
$$g(f(n)) = 2(12n)$$
$$g(f(n)) = 24n$$

11.

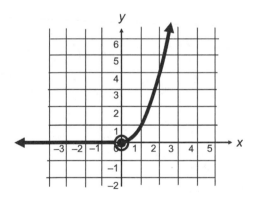

12. The correct answer is shown below

$$g(x) = \begin{cases} 3, & (-\infty, 1] \\ 4, & (1, 3) \\ 5, & (3, \infty) \end{cases}$$

13. The correct answer is 4.

$$f(4) = (4)$$

14. The correct answer is –2.

$$g(-3) = (-3) + 1$$
$$g(-3) = -2$$

15. The correct answer is 25.

$$h(5) = (5)^2$$
$$h(5) = 25$$

Chapter 4

Trigonometry

In this chapter, we'll review the following concepts:

The Pythagorean Theorem
SOHCAHTOA—sine, cosine, and tangent
The unit circle
Inverse trig functions

The Pythagorean Theorem

Trigonometric functions are used often throughout calculus, so it's a good idea to brush up on the basics. We'll start with the Pythagorean Theorem.

The Pythagorean Theorem allows us to calculate the length of one side of a right triangle if we are given the lengths of the other two sides.

The formula for the Pythagorean Theorem is $a^2 + b^2 = c^2$. The letters a and b represent the lengths of the legs of the triangle. The letter c represents the length of the longest side or hypotenuse of the triangle, as shown in the figure below:

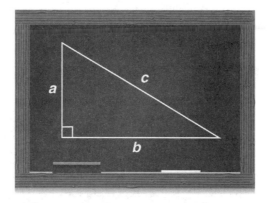

To use the Pythagorean Theorem, plug the given values into the formula and solve for the missing side.

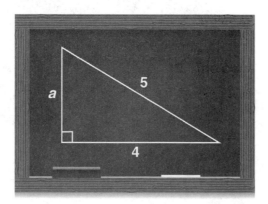

This triangle has one leg that measures 4 units. The length of the hypotenuse is 5 units. Substitute 4 into the formula for the length of one leg of the triangle, b. Substitute 5 into the formula for the length of the hypotenuse, c. Then solve for the length of the missing side, a:

$$a^2 + b^2 = c^2$$
$$a^2 + (4)^2 = (5)^2$$
$$a^2 + 16 = 25$$
$$a^2 = 9$$
$$a = \sqrt{9}$$
$$a = 3$$

The length of the missing side is 3 units.

Practice Questions—The Pythagorean Theorem

Directions: Use the Pythagorean Theorem to answer the questions below. You will find the Practice Question Solutions on page 85.

1. What is the value of *c* in the figure shown?

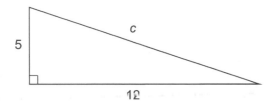

2. What is the value of *c* in the figure shown?

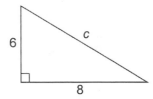

3. What is the value of *a* in the figure shown?

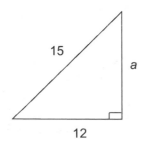

4. What is the length of the hypotenuse in the right triangle shown?

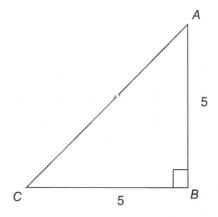

5. What is the value of *b* in the figure shown?

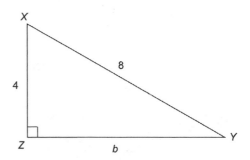

SOHCAHTOA—sine, cosine, and tangent

Trigonometric functions reflect the ratios of certain sides of a right triangle based on their relationship to a given angle, usually denoted by the Greek letter θ.

That's pronounced "theta."

The angle that lies opposite θ is labeled the **opposite** angle. The angle next to θ is labeled the **adjacent** angle. The longest side of the triangle is the hypotenuse:

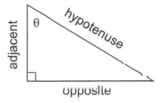

If the location of angle θ shifts, then the names of the sides also change relative to that angle:

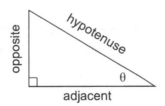

The three main trig functions are sine, cosine, and tangent. Their formulas can be remembered using the phrase **SOHCAHTOA.**

The **sine** of an angle equals the opposite side divided by the hypotenuse:

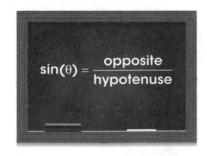

$$\sin(\theta) = \frac{\text{opposite}}{\text{hypotenuse}}$$

The sine formula is represented by the letters **SOH**.

The **cosine** of an angle equals the adjacent side divided by the hypotenuse:

$$\cos(\theta) = \frac{\text{adjacent}}{\text{hypotenuse}}$$

The **tangent** of an angle equals the opposite side divided by the adjacent side:

$$\tan(\theta) = \frac{\text{opposite}}{\text{adjacent}}$$

Cosine is represented by the letters C, A, and H—or **CAH**.

Put it all together and you have **SOHCAHTOA**

Tangent is represented by **TOA**.

In addition to learning the three main trig formulas, it's helpful to also memorize their reciprocals: cosecant, secant, and cotangent. **Cosecant** is the reciprocal of the sine function, **secant** is the reciprocal of the cosine function, and **cotangent** is the reciprocal of the tangent function:

$\csc(\theta) = \dfrac{1}{\sin(\theta)}$	$\sec(\theta) = \dfrac{1}{\cos(\theta)}$	$\cot(\theta) = \dfrac{1}{\tan(\theta)}$

These reciprocal functions can also be represented as follows:

$\csc(\theta) = \dfrac{\text{hypotenuse}}{\text{opposite}}$	$\sec(\theta) = \dfrac{\text{hypotenuse}}{\text{adjacent}}$	$\cot(\theta) = \dfrac{\text{adjacent}}{\text{opposite}}$

In the triangle shown, the side opposite θ measures 4 units. The hypotenuse measures 5 units. To determine the value of $\sin(\theta)$, we plug these side lengths into the sine formula.

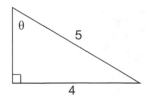

The formula for sine is the opposite side divided by the hypotenuse. Using the formula, plug in the values 4 for the opposite side and 5 for the hypotenuse:

$$\sin(\theta) = \frac{\text{opposite}}{\text{hypotenuse}}$$
$$= \frac{4}{5}$$

The value of $\sin(\theta)$ is $\frac{4}{5}$.

Practice Questions—SOHCAHTOA

Directions: Use SOHCAHTOA to answer the questions below. You will find the Practice Question Solutions on page 86.

6. Find the value of $\sin(\theta)$.

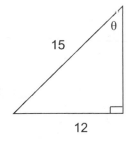

7. Find the value of $\cos(\theta)$.

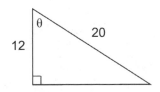

8. Find the value of $\tan(\theta)$.

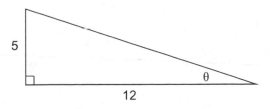

9. Find the value of $\csc(\theta)$.

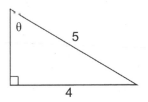

10. Find the value of $\cot(\theta)$.

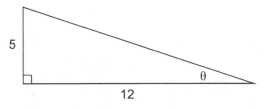

The unit circle

One important tool for determining the value of trig functions is the unit circle. The **unit circle** is a circle drawn on the *x-y* coordinate plane with its center at the origin and a radius of one unit.

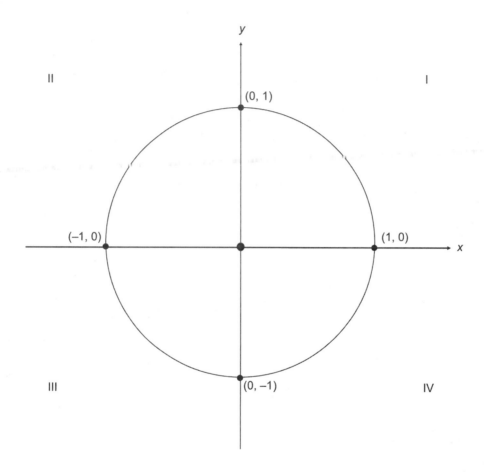

The unit circle is divided into four quadrants, as shown. For the purposes of trigonometry, the unit circle has three distinguishing features. First, lines can be drawn that divide the circle into angles of varying degrees:

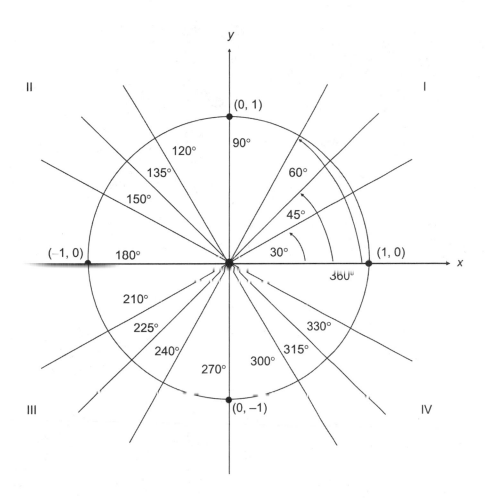

Second, these angle measures can be expressed in a unit of measurement called radians. **Radians** are given in terms of the Greek letter π.

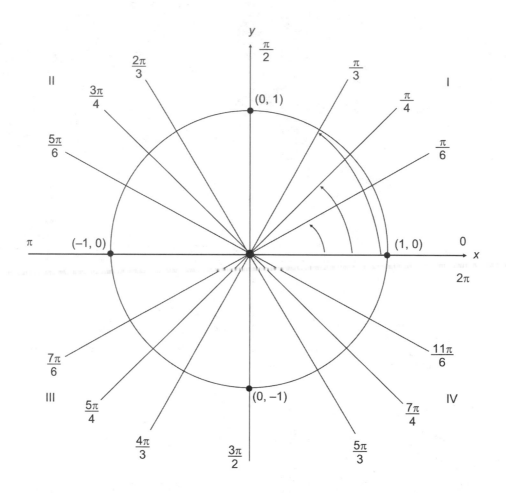

Third, each point at which an angle ray intersects the circle represents the cosine and sine value of the angle. The cosine value is represented by the *x*-coordinate. The sine value is represented by the *y*-coordinate.

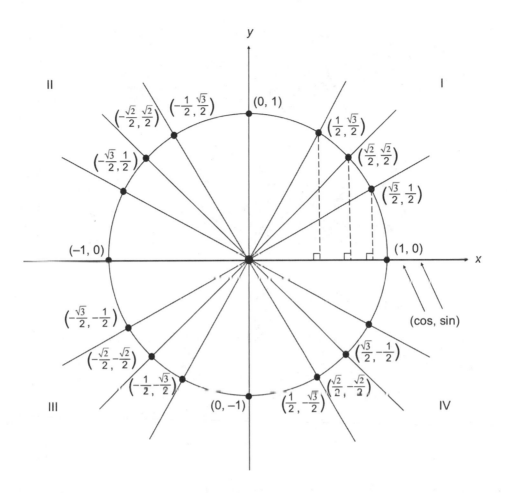

Putting these three features together, and labeling all of the major angles in all four quadrants, produces the following:

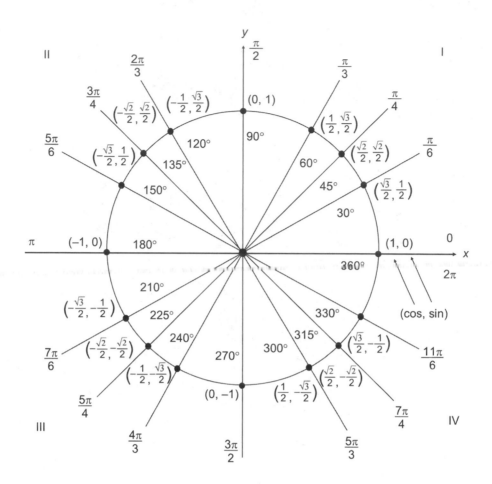

We can use the unit circle to find the value of the trig functions for major angles.

Example

As an example, let's use the unit circle to find the value of $\sin\left(\frac{\pi}{3}\right)$.

Look up $\frac{\pi}{3}$ radians on the unit circle. This corresponds to 60°.
At the point where the angle's top ray intersects the unit circle,

the coordinates are given as $\left(\frac{1}{2}, \frac{\sqrt{3}}{2}\right)$. The y-coordinate is

equal to the sine value of the angle, so $\sin\left(\frac{\pi}{3}\right)$ equals $\frac{\sqrt{3}}{2}$.

Remember that
(*x*, *y*) equals
(cos, sin).

Practice Questions—The unit circle

Directions: Use the unit circle to answer the questions below. You will find the Practice Question Solutions on page 87.

11. What is the value of $\sin\left(\frac{\pi}{6}\right)$?

12. What is the value of $\cos\left(\frac{\pi}{3}\right)$?

13. What is the value of $\sin\left(\frac{\pi}{2}\right)$?

14. What is the value of $\cos\left(\frac{3\pi}{4}\right)$?

15. What is the value of $\sin\left(\frac{4\pi}{3}\right)$?

Inverse trig functions

Inverse trig functions are used to find the measure of an angle of a triangle, given the ratio of the lengths of two sides. If you know the sine value of an angle, for instance, you can find the measurement of that angle using the inverse sine function.

The symbol for an inverse trig function is –1 written as an exponent. The inverse of $\sin(\theta)$ would be written as $\sin^{-1}(\theta)$.

For most common inverse trig functions, we can use the unit circle and work backwards.

What is the value of $\cos^{-1}\left(\frac{\sqrt{3}}{2}\right)$ in quadrant I?

To answer this question, we consult the unit circle. The x-coordinate of a point on the unit circle is equal to its cosine value. The point with an x-coordinate of $\frac{\sqrt{3}}{2}$ in quadrant I is 30°, or $\frac{\pi}{6}$ radians. The measure of the angle can be expressed in either degrees or radians.

Practice Questions—Inverse trig functions

Directions: Use the unit circle to answer the questions below. You will find the Practice Question Solutions on page 87.

16. Find the value of $\sin^{-1}\left(\dfrac{1}{2}\right)$ in quadrant I.

    Express your answer in radians.

17. Find the value of $\cos^{-1}\left(\dfrac{1}{2}\right)$ in quadrant I.

    Express your answer in degrees.

18. Find the value of $\sin^{-1}\left(\dfrac{\sqrt{3}}{2}\right)$ in quadrant I.

    Express your answer in degrees.

19. Find the value of $\cos^{-1}\left(\dfrac{\sqrt{2}}{2}\right)$ in quadrant I.

    Express your answer in radians.

20. Find the value of $\cos^{-1}\left(-\dfrac{\sqrt{3}}{2}\right)$ in quadrant II.

    Express your answer in radians.

Chapter Review

Directions: Use your knowledge of the Pythagorean Theorem, trig functions, and the unit circle to answer the questions below. Solutions can be found on page 88.

1. What is the value of *b* in the figure shown?

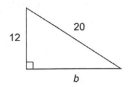

2. What is the length of side *JK* in the right triangle shown?

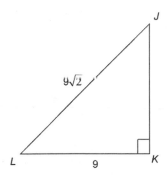

3. What is the value of *c* in the figure shown?

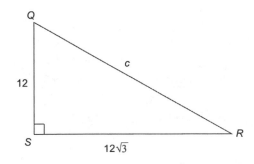

4. Find the value of $\sin(\theta)$.

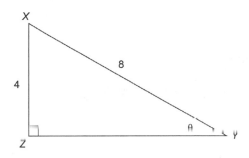

5. Find the value of $\tan(\theta)$.

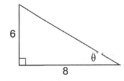

6. Find the value of $\sec(\theta)$.

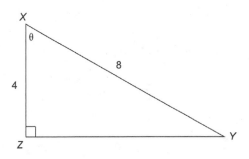

7. Find the value of $\cot(\theta)$.

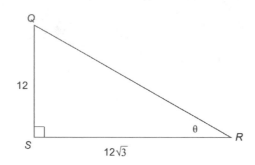

8. What is the value of $\cos\left(\dfrac{4\pi}{3}\right)$?

9. What is the value of $\sin\left(\dfrac{7\pi}{6}\right)$?

10. What is the value of $\cos(2\pi)$?

11. What is the value of $\cos(3\pi)$?

12. Find the value of $\sin^{-1}\left(\dfrac{\sqrt{2}}{2}\right)$ in quadrant II. Express your answer in degrees.

13. Find the value of $\cos^{-1}\left(-\dfrac{1}{2}\right)$ in quadrant III. Express your answer in radians.

14. Find the value of $\sin^{-1}\left(-\dfrac{1}{2}\right)$ in quadrant IV. Express your answer in radians.

15. Find the value of $\cos^{-1}\left(\dfrac{\sqrt{2}}{2}\right)$ in quadrant IV. Express your answer in radians.

Practice Question Solutions

The Pythagorean Theorem

1. The correct answer is 13.

Substitute 5 and 12 for the lengths of the legs of the triangle, a and b. Then solve for the missing length of the hypotenuse, c:

$$a^2 + b^2 = c^2$$
$$(5)^2 + (12)^2 = c^2$$
$$25 + 144 = c^2$$
$$c^2 = 169$$
$$c = \sqrt{169}$$
$$c = 13$$

2. The correct answer is 10.

Substitute 6 and 8 for the lengths of the legs of the triangle, a and b. Then solve for the missing length of the hypotenuse, c:

$$a^2 + b^2 = c^2$$
$$(6)^2 + (8)^2 = c^2$$
$$36 + 64 = c^2$$
$$c^2 = 100$$
$$c = \sqrt{100}$$
$$c = 10$$

3. The correct answer is 9.

In this question, we are given the length of the hypotenuse, c, and the length of one side, b. Substitute 15 for c and 12 for b in the formula. Then solve for the length of the missing side, a:

$$a^2 + b^2 = c^2$$
$$a^2 + (12)^2 = (15)^2$$
$$a^2 + 144 = 225$$
$$a^2 = 81$$
$$a = \sqrt{81}$$
$$a = 9$$

4. The correct answer is $5\sqrt{2}$.

Substitute 5 and 5 for the lengths of the legs of the triangle, a and b. Then solve for the missing length of the hypotenuse, c:

$$a^2 + b^2 = c^2$$
$$(5)^2 + (5)^2 = c^2$$
$$25 + 25 = c^2$$
$$c^2 = 50$$
$$c = \sqrt{50}$$
$$c = \sqrt{25 \times 2}$$
$$c = 5\sqrt{2}$$

5. The correct answer is $4\sqrt{3}$.

Here we are given the length of the hypotenuse, c, and the length of one side, a. Substitute 8 for c and 4 for a in the formula. Then solve for the length of the missing side, b:

$$a^2 + b^2 = c^2$$
$$(4)^2 + b^2 = (8)^2$$
$$16 + b^2 = 64$$
$$b^2 = 48$$
$$b = \sqrt{48}$$
$$b = \sqrt{16 \times 3}$$
$$b = 4\sqrt{3}$$

SOHCAHTOA—sine, cosine, and tangent

6. The correct answer is $\frac{4}{5}$.

The formula for sine is the opposite side divided by the hypotenuse. Using the formula, plug in the values 12 for the opposite side and 15 for the hypotenuse:

$$\sin(\theta) = \frac{\text{opposite}}{\text{hypotenuse}}$$
$$= \frac{12}{15}$$
$$= \frac{4}{5}$$

7. The correct answer is $\frac{3}{5}$.

The formula for cosine is the adjacent side divided by the hypotenuse. Using the formula, plug in the values 12 for the adjacent side and 20 for the hypotenuse:

$$\cos(\theta) = \frac{\text{adjacent}}{\text{hypotenuse}}$$
$$= \frac{12}{20}$$
$$= \frac{3}{5}$$

8. The correct answer is $\frac{5}{12}$.

The formula for tangent is the opposite side divided by the adjacent side. Using the formula, plug in the values 5 for the opposite side and 12 for the adjacent side:

$$\tan(\theta) = \frac{\text{opposite}}{\text{adjacent}}$$
$$= \frac{5}{12}$$

9. The correct answer is $\frac{5}{4}$.

The cosecant function is the reciprocal of the sine function. Its formula is therefore the length of the hypotenuse divided by the length of the opposite side. Using the formula, plug in the values 5 for the hypotenuse and 4 for the opposite side:

$$\csc(\theta) = \frac{\text{hypotenuse}}{\text{opposide}}$$
$$= \frac{5}{4}$$

10. The correct answer is $\frac{12}{5}$.

The cotangent function is the reciprocal of the tangent function. Its formula is therefore the length of the adjacent side divided by the length of the opposite side. Using the formula, plug in the values 12 for the adjacent side and 5 for the opposite side:

$$\cot(\theta) = \frac{\text{adjacent}}{\text{opposide}}$$
$$= \frac{12}{5}$$

The unit circle

11. The correct answer is $\frac{1}{2}$.

    At $\frac{\pi}{6}$, the coordinates on the unit circle are given as $\left(\frac{\sqrt{3}}{2}, \frac{1}{2}\right)$. The y-coordinate is equal to the sine value, so $\sin\left(\frac{\pi}{6}\right)$ equals $\frac{1}{2}$.

12. The correct answer is $\frac{1}{2}$.

    At $\frac{\pi}{3}$, the coordinates on the unit circle are given as $\left(\frac{1}{2}, \frac{\sqrt{3}}{2}\right)$. The x-coordinate is equal to the cosine value, so $\cos\left(\frac{\pi}{3}\right)$ equals $\frac{1}{2}$.

13. The correct answer is 1.

    At $\frac{\pi}{2}$, the coordinates on the unit circle are given as $(0, 1)$. The y-coordinate is equal to the sine value, so $\sin\left(\frac{\pi}{2}\right)$ equals 1.

14. The correct answer is $-\frac{\sqrt{2}}{2}$.

    At $\frac{3\pi}{4}$, the coordinates on the unit circle are given as $\left(-\frac{\sqrt{2}}{2}, \frac{\sqrt{2}}{2}\right)$. The x-coordinate is equal to the cosine value, so $\cos\left(\frac{3\pi}{4}\right)$ equals $-\frac{\sqrt{2}}{2}$.

15. The correct answer is $-\frac{\sqrt{3}}{2}$.

    At $\frac{4\pi}{3}$, the coordinates on the unit circle are given as $\left(-\frac{1}{2}, -\frac{\sqrt{3}}{2}\right)$. The y-coordinate is equal to the sine value, so $\sin\left(\frac{3\pi}{2}\right)$ equals $-\frac{\sqrt{3}}{2}$.

Inverse trig functions

16. The correct answer is $\frac{\pi}{6}$.

    The y-coordinate of a point on the unit circle is equal to the sine value. The point with a y-coordinate of $\frac{1}{2}$ in quadrant I is $\frac{\pi}{6}$. The value of $\sin^{-1}\left(\frac{1}{2}\right)$ expressed in radians is $\frac{\pi}{6}$.

17. The correct answer is 60°.

    The x-coordinate of a point on the unit circle is equal to the cosine value. The point with an x-coordinate of $\frac{1}{2}$ in quadrant I is 60°. The value of $\cos^{-1}\left(\frac{1}{2}\right)$ expressed in degrees is 60°.

18. The correct answer is 60°.

The *y*-coordinate of a point on the unit circle is equal to the sine value. The point with a *y*-coordinate of $\frac{\sqrt{3}}{2}$ in quadrant I is 60°. The value of $\sin^{-1}\left(\frac{\sqrt{3}}{2}\right)$ expressed in degrees is 60°.

19. The correct answer is $\frac{\pi}{4}$.

The *x*-coordinate of a point on the unit circle is equal to the cosine value. The point with an *x*-coordinate of $\frac{\sqrt{2}}{2}$ in quadrant I is $\frac{\pi}{4}$. The value of $\cos^{-1}\left(\frac{\sqrt{2}}{2}\right)$ expressed in radians is $\frac{\pi}{4}$.

20. The correct answer is $\frac{5\pi}{6}$.

The *x*-coordinate of a point on the unit circle is equal to the cosine value. The point with an *x*-coordinate of $-\frac{\sqrt{3}}{2}$ in quadrant II is $\frac{5\pi}{6}$. The value of $\cos^{-1}\left(-\frac{\sqrt{3}}{2}\right)$ expressed in radians is $\frac{5\pi}{6}$.

Chapter Review Solutions

1. The correct answer is 16.

Here we are given the length of the hypotenuse, *c*, and the length of one side, *a*. Substitute 20 for *c* and 12 for *a* in the formula. Then solve for the length of the missing side, *b*:

$$a^2 + b^2 = c^2$$
$$(12)^2 + b^2 = (20)^2$$
$$144 + b^2 = 400$$
$$b^2 = 256$$
$$b = \sqrt{256}$$
$$b = 16$$

2. The correct answer is 9.

In this question, we are given the length of the hypotenuse and the length of one side. Let the length of side *JK* be represented by the variable *a*. Substitute $9\sqrt{2}$ for *c* and 9 for *b* in the formula. Then solve for the length of the missing side, *a*:

$$a^2 + b^2 = c^2$$
$$a^2 + (9)^2 = \left(9\sqrt{2}\right)^2$$
$$a^2 + 81 = 81 \times 2$$
$$a^2 + 81 = 162$$
$$a^2 = 81$$
$$a = 9$$

3. The correct answer is 24

Substitute 12 and $12\sqrt{3}$ for the lengths of the legs of the triangle, a and b. Then solve for the missing length of the hypotenuse, c:

$$a^2 + b^2 = c^2$$
$$(12)^2 + (12\sqrt{3})^2 = c^2$$
$$144 + (144 \times 3) = c^2$$
$$144 + 432 = c^2$$
$$c^2 = 576$$
$$c = \sqrt{576}$$
$$c = 24$$

4. The correct answer is $\dfrac{1}{2}$

The formula for sine is the opposite side divided by the hypotenuse. Using the formula, plug in the values 4 for the opposite side and 8 for the hypotenuse:

$$\sin(\theta) = \frac{\text{opposite}}{\text{hypotenuse}}$$
$$= \frac{4}{8}$$
$$= \frac{1}{2}$$

5. The correct answer is $\dfrac{3}{4}$.

The formula for tangent is the opposite side divided by the adjacent side. Using the formula, plug in the values 6 for the opposite side and 8 for the adjacent side:

$$\tan(\theta) = \frac{\text{opposite}}{\text{adjacent}}$$
$$= \frac{6}{8}$$
$$= \frac{3}{4}$$

6. The correct answer is 2.

The secant function is the reciprocal of the cosine function. Its formula is therefore the length of the hypotenuse divided by the length of the adjacent side. Using the formula, plug in the values 8 for the hypotenuse and 4 for the adjacent side:

$$\sec(\theta) = \frac{\text{hypotenuse}}{\text{adjacent}}$$
$$= \frac{8}{4}$$
$$= 2$$

7.

The correct answer is $\sqrt{3}$.

The cotangent function is the reciprocal of the tangent function. Its formula is therefore the length of the adjacent side divided by the length of the opposite side. Using the formula, plug in the values $12\sqrt{3}$ for the adjacent side and 12 for the opposite side:

$$\cot(\theta) = \frac{\text{adjacent}}{\text{opposide}}$$
$$= \frac{12\sqrt{3}}{12}$$
$$= \sqrt{3}$$

8. The correct answer is $-\frac{1}{2}$.

At $\frac{4\pi}{3}$, the coordinates on the unit circle are given as $\left(-\frac{1}{2}, -\frac{\sqrt{3}}{2}\right)$. The x-coordinate is equal to the cosine value, so $\cos\left(\frac{4\pi}{3}\right)$ equals $-\frac{1}{2}$.

9. The correct answer is $-\frac{1}{2}$.

At $\frac{7\pi}{6}$, the coordinates on the unit circle are given as $\left(-\frac{\sqrt{3}}{2}, -\frac{1}{2}\right)$. The y-coordinate is equal to the sine value, so $\sin\left(\frac{7\pi}{6}\right)$ equals $-\frac{1}{2}$.

10. The correct answer is 1.

The value of 2π radians represents $360°$ on the unit circle. At 2π, the coordinates on the unit circle are the same as the coordinates for 0 radians: $(1, 0)$. The x-coordinate is equal to the cosine value, so $\cos(2\pi)$ equals 1.

11. The correct answer is –1.

The value of 3π radians represents one and one-half turns around the unit circle. This is equivalent to $360°$ plus another $180°$, or $540°$. At 3π, the coordinates on the unit circle are the same as the coordinates for π radians: $(-1, 0)$. The x-coordinate is equal to the cosine value, so $\cos(3\pi)$ equals –1.

12. The correct answer is $135°$.

The values of inverse trig functions are angle measures.

The y-coordinate of a point on the unit circle is equal to the sine value. The point with a y-coordinate of $\frac{\sqrt{2}}{2}$ in quadrant II is $135°$. The value of $\sin^{-1}\left(\frac{\sqrt{2}}{2}\right)$ expressed in degrees is $135°$.

13. The correct answer is $\frac{4\pi}{3}$.

The x-coordinate of a point on the unit circle is equal to the cosine value. The point with an x-coordinate of $-\frac{1}{2}$ in quadrant III is $\frac{4\pi}{3}$. The value of $\cos^{-1}\left(-\frac{1}{2}\right)$ expressed in radians is $\frac{4\pi}{3}$.

14. The correct answer is $\frac{11\pi}{6}$.

The y-coordinate of a point on the unit circle is equal to the sine value. The point with a y-coordinate of $-\frac{1}{2}$ in quadrant IV is $\frac{11\pi}{6}$. The value of $\sin^{-1}\left(-\frac{1}{2}\right)$ expressed in radians is $\frac{11\pi}{6}$.

15. The correct answer is $\frac{7\pi}{4}$.

The *x*-coordinate of a point on the unit circle is equal to the cosine value. The point with an *x*-coordinate of $\frac{\sqrt{2}}{2}$ in quadrant IV is $\frac{7\pi}{4}$. The value of $\cos^{-1}\left(\frac{\sqrt{2}}{2}\right)$ expressed in radians is $\frac{7\pi}{4}$.

Chapter 5

Limits

In this chapter, we'll review the following concepts:

What is a limit?
Properties of limits
Taking limits
Infinite limits

What is a limit?

The limit process is the most fundamental tool in your calculus toolkit. It is an important part of what defines the derivative and the integral, which are the two main operations performed in calculus, as we'll see in later chapters.

When mathematicians use the word "limit," this term makes it sound as if a limit is a constraint or some type of limitation. But a limit is not a constraint, as the name would suggest. Instead, a limit is a value determined based on trends in a given set of data.

A limit is a trending value.

On the graph below, we can see that as we are approaching $x = 4$ from values less than 4, as we get closer and closer to $x = 4$, we approach a certain y-value, 3. As we approach x from the other direction, from values greater than 4, we also approach the y-value of 3. If the limit of the function exists, the function will approach the same y-value as we approach x from both directions.

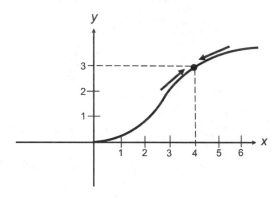

The limit is the value that y gets close to when we get closer and closer to x from the right and left sides of the graph. In this case, the limit of the function as we get closer to 4 is 3. Based on the trends we see in the graph, when we approach $x = 4$ from either side, the y-value of the function approaches 3.

It's important to understand that the limit is not the same as the y-value of the function. The limit reflects what the y-value *would* be if the behavior that we're observing continued as expected. In reality, sometimes this observed behavior does continue as expected, and sometimes it does not. In the example above, the behavior that we're observing actually does follow the observed trend, because we have a real point at (4, 3). So in this case, the limit and the y-value of the function are the same number.

This is not always the case, however. The *y*-value of the function might be something different than the limit altogether. In the graph below, for instance, the limit of the function as *x* approaches 4 is still 3. As we approach *x* = 4 from both directions on the graph, we also approach the *y*-value of 3.

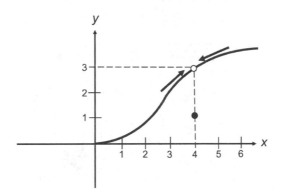

This time, however, the trend that we're observing doesn't play out on the graph of the function, because there is a gap in the curve at the point *x* = 4. Instead of following the trend and continuing as expected, the function changes unexpectedly at *x* = 4 and has a *y*-value of 1. So the *function* is defined differently at that value of *x*, but it still has the same limit as the function shown in the first graph. At the point *x* = 4, the actual *y*-value of this function is 1, while the limit of the function as *x* approaches 4 is still 3.

When dealing with a limit, we are concerned with the conclusion that comes from observing trends in the behavior of the function. We are not concerned with the value of the function itself. If the function were to continue to follow the trend shown on the curve as we approach *x* from both directions, it would get closer and closer to a specific *y*-value. This is the limit.

Properties of limits

The limit of a function $f(x)$ can be expressed as follows. The letter L is used to represent the limit of the function $f(x)$ as the value of x approaches some number, c. In limit notation, we would write that as:

$$\lim_{x \to c} f(x) = L$$

This just means that as the value of x approaches some number, c, the limit of the function $f(x)$ is some number, L.

There are a few properties of limits that can be helpful to know. To identify these properties, we will consider two functions, $f(x)$ and $g(x)$. We will use the letters c and k to represent constants.

1. The limit of a constant is itself.

$$\lim_{x \to c} f(k) = k$$

2. Given a function such as $f(x) = x$, where the output value equals the input value, the limit of the function is the input value.

$$\lim_{x \to c} f(x) = c$$

3. The limits of functions can be added together.

$$\lim_{x \to c} \left(f(x) + g(x) \right) = \left(\lim_{x \to c} f(x) \right) + \left(\lim_{x \to c} g(x) \right)$$

4. The limits of functions can be subtracted from each other.

$$\lim_{x \to c} \left(f(x) - g(x) \right) = \left(\lim_{x \to c} f(x) \right) - \left(\lim_{x \to c} g(x) \right)$$

5. The limits of functions can be multiplied.

$$\lim_{x \to c} \left(f(x) \cdot g(x) \right) = \left(\lim_{x \to c} f(x) \right) \left(\lim_{x \to c} g(x) \right)$$

6. The limits of functions can be divided, as long as the divisor does not equal 0.

$$\lim_{x \to c} \left(\frac{f(x)}{g(x)} \right) = \frac{\left(\lim_{x \to c} f(x) \right)}{\left(\lim_{x \to c} g(x) \right)}$$

Taking limits

Now that we've gained an understanding of limits and their properties, we can use certain techniques for finding the limits of specific functions. There are four types of limits that we'll look at next: polynomials, rational functions, composite functions, and trig functions. Here are the processes for taking the limit of each.

Limits of polynomials

To take the limits of polynomials, we substitute the value of x into the polynomial and simplify the expression.

To take this limit, we substitute 2 for x and solve:

$$\lim_{x \to 2}\left(2x^2 + 1\right) = \left(2(2)^2 + 1\right)$$
$$= \left(2(4) + 1\right)$$
$$= (8 + 1)$$
$$= 9$$

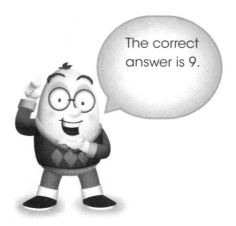

The correct answer is 9.

In these cases, the function does exist at the value of x where we are trying to find the limit. We can therefore try to simply plug in the x-value and solve for the y-value of the function. As long as the result is not a number divided by zero, the result will be the limit.

Limits of rational functions

Rational functions are functions that can be expressed as ratios, which means they can be written as fractions. For these functions, we take the limit just as we did with polynomials above.

$$\lim_{x \to -2} \frac{x + 6}{x^2}$$

For this problem, we substitute 2 in for x and solve:

$$\lim_{x \to -2} \frac{x + 6}{x^2} = \frac{(-2) + 6}{(-2)^2}$$
$$= \frac{4}{4}$$
$$= 1$$

The limit of the rational function is 1.

When dealing with rational functions, sometimes the process of substitution results in a value of $\frac{0}{0}$. In these cases, we use the **cancellation technique** instead to obtain the result.

Try this problem:

$$\lim_{x \to 2} \frac{x^2 - 4}{x - 2}.$$

First, we try to substitute 2 for x and solve:

$$\lim_{x \to 2} \frac{x^2 - 4}{x - 2} = \frac{(2)^2 - 4}{(2) - 2}$$

$$= \frac{4 - 4}{2 - 2}$$

$$= \frac{0}{0}$$

The value $\frac{0}{0}$ is an indeterminate form, meaning we cannot find the limit by using substitution alone. If we factor the numerator, we can see that $x - 2$ is a common term in both the numerator and the denominator. So, we can use the cancellation technique to find the limit.

$$\lim_{x \to 2} \frac{x^2 - 4}{x - 2} = \frac{(x - 2)(x + 2)}{x - 2}$$

$$= x + 2$$

$$= (2) + 2$$

$$= 4$$

The correct answer is 4.

Limits of composite functions

To find the limits of composite functions, we first perform the operation that is represented by the composite function. Then substitute in the value of x and solve. This process is the same as that used for finding the limit of a polynomial, but the operation of the composite function is performed first.

$$f(x) = x^2, g(x) = x + 1$$
$$\lim_{x \to 1} g(f(x))$$

Here we are asked to find the limit of the function $g(f(x))$, as x approaches 1. First, we substitute x^2 for x in $g(x)$ to obtain the composite function:

$$g(f(x)) = g(x^2)$$

$$= x^2 + 1$$

Now we can find the limit by substituting 1 for x:

$$\lim_{x \to 1} g\left(f(x)\right) = \lim_{x \to 1} x^2 + 1$$
$$= (1)^2 + 1$$
$$= 2$$

The correct answer is 2.

Limits of trig functions

To take the limits of trig functions, we use the unit circle.

$$\lim_{x \to \frac{\pi}{6}} \sin(x)$$

To solve this problem, first substitute $\frac{\pi}{6}$ for x. Then consult the unit circle to determine the value of $\sin\left(\frac{\pi}{6}\right)$.

$$\lim_{x \to \frac{\pi}{6}} \sin(x) = \sin\left(\frac{\pi}{6}\right)$$

$$= \frac{1}{2}$$

The value of $\sin\left(\frac{\pi}{6}\right)$ is $\frac{1}{2}$, so the limit of $\sin(x)$ as x approaches $\frac{\pi}{6}$ is $\frac{1}{2}$.

Practice Questions—Taking limits

Directions: Find the limit of the polynomials, rational functions, composite functions, and trig functions shown. You will find the Practice Question Solutions on page 111.

Find the limit of the polynomial.

1. $\lim\limits_{x \to 1} 2x^5$

2. $\lim\limits_{x \to -2} \left(-5x^3 + x - 1\right)$

Find the limit of the rational function.

3. $\lim\limits_{x \to 3} \dfrac{-x^3 + 1}{2x^4}$

Find the limit of the composite function.

4. $f(x) = x^2, g(x) = 3x^3$

$\lim\limits_{x \to -1} f(g(x))$

Find the limit of the trigonometric function.

5. $\lim\limits_{x \to \frac{5\pi}{6}} \sin(x)$

Infinite limits

In the definition of a limit given earlier, a limit exists when the function approaches a certain *y*-value as the *x*-value is approached from both sides of the graph. It is also possible, however, to have cases where a limit does not exist. We will look at two cases in this chapter.

First, when the graph of the function does not approach the same *y*-value as we approach *x* from the left and from the right, the limit does not exist. The graph below is an example of such a function:

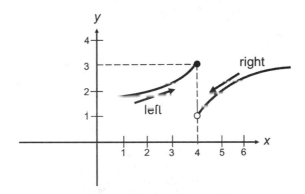

As we approach *x* = 4 from the right, the *y*-value of the function approaches 1. As we approach *x* = 4 from the left, the *y*-value approaches 3. So, the limit of this function as *x* approaches 4 from the right is 1. The limit of the function as *x* approaches 4 from the left is 3.

These two limits are not equal. So, for this function, we say that as *x* approaches 4, the regular two-sided limit described above does not exist. Instead, we call these **one-sided limits**, and we denote them as follows:

$\lim_{x \to 4^+} f(x)$	$\lim_{x \to 4^-} f(x)$
The limit as x approaches 4 from the right	The limit as x approaches 4 from the left

The + means from the right, and the – means from the left.

Also, when you take a limit and the result is a number divided by zero, the limit does not exist at that point of the function. If taking the limit produces a number with a constant as the numerator and 0 as the denominator, as shown below, we say the limit is infinity:

In this case, the value of the numerator is fixed. As the denominator becomes smaller and smaller, the value of the fraction becomes larger and larger. The value of the fraction trends toward infinity.

If the numerator is a negative constant and the denominator is 0, the limit is negative infinity:

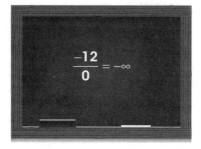

The graph below shows an example of a function with one-sided infinite limits. As x approaches 3 from the right, the y-value approaches positive infinity. As x approaches 3 from the left, the y-value approaches negative infinity:

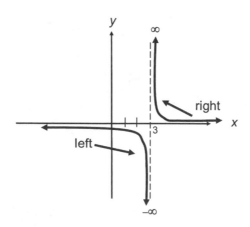

The dotted line at $x = 3$ is known as a **vertical asymptote.** A vertical asymptote shows a value of x for which the function is undefined. As we get closer to that value of x, the y-value gets closer to positive or negative infinity.

Whenever we take a limit that results in a constant, c, divided by 0, there will be a vertical asymptote at $x = c$. If the result is $\frac{2}{0}$, for instance, there will be a vertical asymptote at $x = 2$.

One-sided infinite limits

To solve problems involving one-sided infinite limits, we start with substitution. If the result is a constant divided by 0, this tells us that the graph has a vertical asymptote. Once we graph the function with the indicated asymptote, we can determine its one-sided limits.

For example, to solve the problem below, we must find the limit as x approaches 2 from the left and from the right.

$$\lim_{x \to -2} \frac{5x^2}{x + 2}$$

To find the limit, we first use direct substitution. Substitute –2 into the rational function for x:

$$\lim_{x \to -2} \frac{5x^2}{x + 2} = \frac{5(-2)^2}{(-2) + 2}$$
$$= \frac{20}{0}$$

Since the denominator equals zero and the numerator does not equal zero, there is an asymptote at $x = -2$. Graphing the function allows us to determine the one-sided limits:

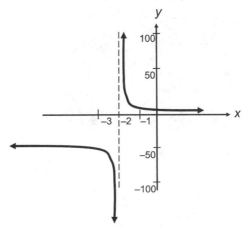

Note: Figure not drawn to scale.

The graph shows us that from the left, the limit is negative infinity:

$$\lim_{x \to -2^-} \frac{5x^2}{x+2} = -\infty$$

We can also see that from the right, the limit of the function is positive infinity:

$$\lim_{x \to -2^+} \frac{5x^2}{x+2} = \infty$$

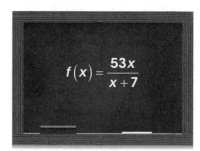

The correct answer is from the left $-\infty$, and from the right ∞.

Finding vertical asymptotes

We can find the vertical asymptotes of a function without graphing by using algebra. Consider the rational function shown:

$$f(x) = \frac{53x}{x+7}$$

Here we are given a rational function with the expression $x + 7$ in the denominator. To find the vertical asymptotes, we are looking for values of x that will result in the denominator equaling zero and the numerator not equaling zero. We can see that when $x = -7$, the function does result in such a value:

$$f(x) = \frac{53x}{x + 7}$$
$$= \frac{53(-7)}{(-7) + 7}$$
$$= \frac{-371}{0}$$

Since the denominator equals zero but the numerator does not equal zero when $x = -7$, there is an asymptote at $x = -7$.

Properties of infinite limits

The properties of infinite limits can be used to find limits that represent sums, differences, products, and quotients of individual limits. The following properties can be helpful to know:

Sum property	$\infty + \text{constant} = \infty$
Difference property	$\infty - \text{constant} = \infty$
Product property	$\infty \times \text{constant} = \infty$ $\infty \times -\text{constant} = -\infty$
Quotient property	$\dfrac{\infty}{\text{constant}} = 0$

Here is an example of how the properties can be used in solving problems. Find the indicated limit:

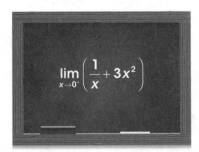

$$\lim_{x \to 0^+} \left(\frac{1}{x} + 3x^2 \right)$$

Since this limit consists of a sum of two functions, we examine one function at a time. We can see that when $x = 0$, the first function's denominator equals zero while its numerator does not equal zero. We can conclude that there is a vertical asymptote at $x = 0$. By graphing the first function, we can find its one-sided limit.

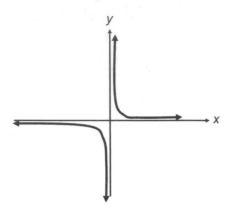

$$\lim_{x \to 0^+} \left(\frac{1}{x} \right) = \infty$$

The limit of the second function equals a constant:

$$\lim_{x \to 0^+} \left(3x^2 \right) = 0$$

Using the sum property of infinite limits, we can see:

$$\lim_{x \to 0^+} \left(\frac{1}{x} + 3x^2 \right) = \left(\lim_{x \to 0^+} \frac{1}{x} \right) + \left(\lim_{x \to 0^+} 3x^2 \right)$$

$$= \infty + 0$$

$$= \infty$$

The correct answer is ∞.

Practice Questions—Infinite limits

Directions: Solve the problems below according to the instructions given. You will find the Practice Question Solutions on page 112.

Find the limit as the function approaches each value of x from the left and from the right.

6. $\lim\limits_{x \to -1} \dfrac{6x + 5}{(x + 1)^2}$

7. $\lim\limits_{x \to 3} \dfrac{6}{(x - 3)^3}$

Find all of the vertical asymptotes of the function.

8. $g(x) = \dfrac{x^2 + 7x + 10}{x^3 + 7x^2 + 4x - 12}$

Find the indicated limit.

9. $\lim\limits_{x \to 1^+} \left(\dfrac{2x}{x^2 - 1} - 6 \right)$

10. $\lim\limits_{x \to 4^-} \dfrac{7x^2 + 2x + 1}{\dfrac{1}{4 - x}}$

Chapter Review

Directions: Use your knowledge of taking limits and infinite limits to answer the questions below. Solutions can be found on page 115.

Find the limit of the polynomials and rational functions.

1. $\lim\limits_{x \to 4} 4x^4$

2. $\lim\limits_{x \to -3} \left(3x^4 - 2x^3 - 8x - 300\right)$

3. $\lim\limits_{x \to -\sqrt{2}} \dfrac{x^4 - 5}{2x^2}$

4. $\lim\limits_{x \to -2} \dfrac{3x^2 - 6x - 24}{x + 2}$

5. Find the limit of the composite function.

$$h(x) = \frac{3}{x^2}, \, j(x) = \frac{2x + 5}{-x^3 - x^2 + x}$$
$$\lim\limits_{x \to 1} j\big(h(x)\big)$$

Find the limit of the trigonometric functions.

6. $\lim\limits_{x \to \frac{\pi}{2}} \cos(2x)$

7. $\lim\limits_{x \to \frac{\pi}{3}} \left(\csc(2x) - \cos(x) + \tan(4x)\right)$

Find the limit as the function approaches each value of x from the left and from the right.

8. $\lim\limits_{x \to 7} \dfrac{x^2 - 7}{x - 7}$

9. $\lim\limits_{x \to \frac{1}{2}} \dfrac{-3x^2 + 3}{2x - 1}$

10. $\lim\limits_{x \to \frac{\sqrt{2}}{2}} \dfrac{6x - 5}{x^2 - \frac{1}{2}}$

Find all of the vertical asymptotes of each function.

11. $g(x) = \dfrac{4x^3 + 3x^2 + 2x - 1}{x - 1}$

12. $f(x) = \dfrac{x^3 + x^2 - 9x - 9}{x^2 - 9}$

Find the indicated limit.

13. $\lim\limits_{x \to 2^+} \left(\dfrac{3}{2x - 4} + 12x^4\right)$

14. $g(x) = \dfrac{4}{(x - 1)^3}, \, h(x) = -2x^2 - 3$
$$\lim\limits_{x \to 1^+} \big(g(x)h(x)\big)$$

15. $\lim\limits_{x \to 2^+} \dfrac{x^2 - 1}{\left(\dfrac{1}{x - 2}\right)}$

Practice Question Solutions

Taking limits

1. The correct answer is 2.

Substitute 1 for x and solve.

$$\lim_{x \to 1} 2x^5 = 2(1)^5$$
$$= 2$$

2.

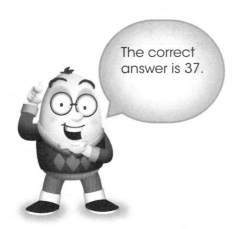

The correct answer is 37.

Substitute –2 for x and solve:

$$\lim_{x \to -2} \left(-5x^3 + x - 1\right) = \left(-5(-2)^3 + (-2) - 1\right)$$
$$= 37$$

3. The correct answer is $-\dfrac{13}{81}$.

Substitute 3 for x and solve:

$$\lim_{x \to 3} \frac{-x^3 + 1}{2x^4} = \frac{-(3)^3 + 1}{2(3)^4}$$
$$= \frac{-26}{162}$$
$$= -\frac{13}{81}$$

4. The correct answer is 9.

First we substitute $3x^3$ for x in $f(x)$ to obtain the composite function:

$$f\left(g(x)\right) = f\left(3x^3\right)$$
$$= \left(3x^3\right)^2$$

Now we can find the limit by substituting –1 for x:

$$\lim_{x \to -1} f\left(g(x)\right) = \lim_{x \to -1} \left(3x^3\right)^2$$
$$= \left(3(-1)^3\right)^2$$
$$= (-3)^2$$
$$= 9$$

5. The correct answer is $\dfrac{1}{2}$.

Substitute $\dfrac{5\pi}{6}$ for x and solve. Use the unit circle if necessary so we can get an exact answer:

$$\lim_{x \to \frac{5\pi}{6}} \sin(x) = \sin\left(\frac{5\pi}{6}\right)$$
$$= \frac{1}{2}$$

Infinite limits

6. The correct answer is from the left $-\infty$, from the right $-\infty$.

$$\lim_{x \to -1} \frac{6x+5}{(x+1)^2} = \frac{6(-1)+5}{((-1)+1)^2}$$

$$= \frac{-1}{0}$$

Since the denominator equals zero and the numerator does not equal zero, there is an asymptote at $x = -1$. Graphing the function allows us to determine the one-sided limits.

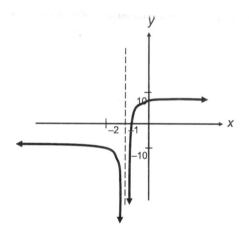

Note: Figure not drawn to scale.

In addition, you may need to change the window on your graphing calculator to view the graph properly.

From the left:

$$\lim_{x \to -1^-} \frac{6x+5}{(x+1)^2} = -\infty$$

From the right:

$$\lim_{x \to -1^+} \frac{6x+5}{(x+1)^2} = -\infty$$

7. The correct answer is from the left $-\infty$, from the right ∞.

$$\lim_{x \to 3} \frac{6}{(x-3)^3} = \frac{6}{((3)-3)^3}$$

$$= \frac{6}{0}$$

Since the denominator equals zero and the numerator does not equal zero, there is an asymptote at $x = 3$. Graphing the function allows us to determine the one-sided limits.

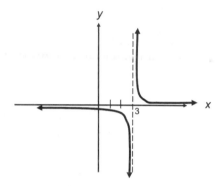

From the left:

$$\lim_{x \to 3^-} \frac{6}{(x-3)^3} = -\infty$$

From the right:

$$\lim_{x \to 3^+} \frac{6}{(x-3)^3} = \infty$$

8. The correct answer is $x = 1$ and $x = -6$.

To find the asymptotes, we are looking for values of x that will result in the denominator equaling zero and the numerator not equaling zero. To make our job easier, we need to factor the numerator and the denominator:

$$g(x) = \frac{x^2 + 7x + 10}{x^3 + 7x^2 + 4x - 12}$$

$$= \frac{(x+2)(x+5)}{(x-1)(x+6)(x+2)}$$

The $x + 2$ term can cancel in the numerator and denominator, so $x = -2$ is not an asymptote. When $x = 1$, the function is:

$$g(1) = \frac{((1)+5)}{((1)-1)((1)+6)}$$

$$= \frac{6}{0}$$

And when $x = -6$, the function is:

$$g(-6) = \frac{((-6)+5)}{((-6)-1)((-6)+6)}$$

$$= \frac{-1}{0}$$

Since the denominator equals zero but the numerator does not equal zero when $x = 1$ or when $x = -6$, there are asymptotes at $x = 1$ and $x = -6$.

9. The correct answer is ∞.

Since this limit consists of a difference of two functions, we examine one function at a time. We can see that when $x = 1$, the first function's denominator equals zero while its numerator does not equal zero. We can conclude that there is a vertical asymptote at $x = 1$. (There is also a vertical asymptote at $x = -1$, but that does not affect the limit we are looking for.) By graphing the first function, we can find its one-sided limit.

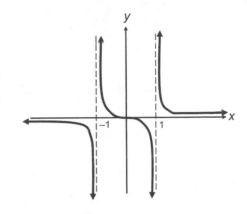

$$\lim_{x \to 1^+} \left(\frac{2x}{x^2 - 1} \right) = \infty$$

The limit of the second function equals a constant.

$$\lim_{x \to 1^+} (6) = 6$$

Using the difference property of infinite limits, we can see:

$$\lim_{x \to 1^+} \left(\frac{2x}{x^2 - 1} - 6 \right) = \infty$$

10. The correct answer is 0.

Since this limit consists of a quotient of two functions, we examine one function at a time. We can see that when $x = 4$, the function in the numerator equals a constant.

$$\lim_{x \to 4^-}\left(7x^2 + 2x + 1\right) = 7(4)^2 + 2(4) + 1$$
$$= 121$$

We can see that when $x = 4$, the denominator of the second function equals zero while its numerator does not equal zero. We can conclude that there is a vertical asymptote at $x = 4$. By graphing the second function, we can find its one-sided limit.

$$\lim_{x \to 4^-} \frac{1}{4 - x} = \infty$$

Using the quotient property of infinite limits, we can see:

$$\lim_{x \to 4^-} \frac{7x^2 + 2x + 1}{\dfrac{1}{4 - x}} = \frac{121}{\infty}$$
$$= 0$$

Chapter Review Solutions

1. The correct answer is 1024.

   Substitute 4 for x and solve:

   $$\lim_{x \to 4} 4x^4 = 4(4)^4$$
   $$= 1024$$

2. The correct answer is 21.

   Substitute –3 for x and solve:

   $$\lim_{x \to -3} \left(3x^4 - 2x^3 - 8x - 300\right) = 3(-3)^4 - 2(-3)^3 - 8(-3) - 300$$
   $$= 243 + 54 + 24 - 300$$
   $$= 21$$

3. The correct answer is $-\dfrac{1}{4}$.

   Substitute $-\sqrt{2}$ for x and solve:

   $$\lim_{x \to -\sqrt{2}} \frac{x^4 - 5}{2x^2} = \frac{\left(-\sqrt{2}\right)^4 - 5}{2\left(-\sqrt{2}\right)^2}$$
   $$= \frac{4 - 5}{2(2)}$$
   $$= -\frac{1}{4}$$

4. First we try to substitute –2 for x and solve:

   $$\lim_{x \to -2} \frac{3x^2 - 6x - 24}{x + 2} = \frac{3(-2)^2 - 6(-2) - 24}{(-2) + 2}$$
   $$= \frac{0}{0}$$

The value $\frac{0}{0}$ is an indeterminate form, meaning we cannot find the limit by using substitution alone. If we factor the numerator, we can see that $x + 2$ is a common term in both the numerator and the denominator. So we can use the cancellation technique to find the limit.

$$\lim_{x \to -2} \frac{3x^2 - 6x - 24}{x + 2} = \frac{3(x + 2)(x - 4)}{x + 2}$$
$$= 3(x - 4)$$
$$= 3((-2) - 4)$$
$$= -18$$

5. The correct answer is $-\frac{1}{3}$.

First we substitute $\frac{3}{x^2}$ for x in $j(x)$ to obtain the composite function.

$$j(h(x)) = \frac{2\left(\dfrac{3}{x^2}\right) + 5}{-\left(\dfrac{3}{x^2}\right)^3 - \left(\dfrac{3}{x^2}\right)^2 + \left(\dfrac{3}{x^2}\right)}$$

Now we can find the limit by substituting 1 for x:

$$\lim_{x \to 1} j(h(x)) = \lim_{x \to 1} \frac{2\left(\dfrac{3}{x^2}\right) + 5}{-\left(\dfrac{3}{x^2}\right)^3 - \left(\dfrac{3}{x^2}\right)^2 + \left(\dfrac{3}{x^2}\right)}$$

$$= \frac{2\left(\dfrac{3}{(1)^2}\right) + 5}{-\left(\dfrac{3}{(1)^2}\right)^3 - \left(\dfrac{3}{(1)^2}\right)^2 + \left(\dfrac{3}{(1)^2}\right)}$$

$$= -\frac{11}{33}$$

$$= -\frac{1}{3}$$

6. The correct answer is –1.

Substitute $\frac{\pi}{2}$ for x and solve, using the unit circle if necessary so we can get an exact answer:

$$\lim_{x \to \frac{\pi}{2}} \cos(2x) = \cos\left(2\left(\frac{\pi}{2}\right)\right)$$
$$= \cos(\pi)$$
$$= -1$$

7. The correct answer is $\frac{5\sqrt{3}}{3} - \frac{1}{2}$.

Substitute $\frac{\pi}{3}$ for x and solve. Use the unit circle if necessary so we can get an exact answer:

$$\lim_{x \to \frac{\pi}{3}} \left(\csc(2x) - \cos(x) + \tan(4x) \right) = \csc\left(2\left(\frac{\pi}{3}\right) \right) - \cos\left(\frac{\pi}{3}\right) + \tan\left(4\left(\frac{\pi}{3}\right) \right)$$

$$= \frac{1}{\sin\left(\frac{2\pi}{3}\right)} - \cos\left(\frac{\pi}{3}\right) + \frac{\sin\left(\frac{4\pi}{3}\right)}{\cos\left(\frac{4\pi}{3}\right)}$$

$$= \frac{1}{\left(\frac{\sqrt{3}}{2}\right)} - \frac{1}{2} + \frac{\left(-\frac{\sqrt{3}}{2}\right)}{\left(-\frac{1}{2}\right)}$$

$$= \frac{2}{\sqrt{3}} - \frac{1}{2} + \sqrt{3}$$

$$= \frac{2}{\sqrt{3}}\left(\frac{\sqrt{3}}{\sqrt{3}}\right) - \frac{1}{2} + \sqrt{3}$$

$$= \frac{2\sqrt{3}}{3} - \frac{1}{2} + \sqrt{3}$$

$$= \frac{5\sqrt{3}}{3} - \frac{1}{2}$$

8. The correct answer is from the left $-\infty$, from the right ∞.

$$\lim_{x \to 7} \frac{x^2 - 7}{x - 7} = \frac{(7)^2 - 7}{(7) - 7}$$

$$= \frac{42}{0}$$

Since the denominator equals zero and the numerator does not equal zero, there is an asymptote at $x = 7$. Graphing the function allows us to determine the one-sided limits.

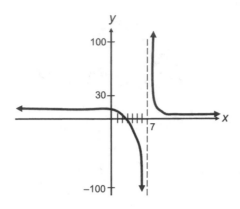

Note: Figure not drawn to scale.

From the left:

$$\lim_{x \to 7^-} \frac{x^2 - 7}{x - 7} = -\infty$$

From the right:

$$\lim_{x \to 7^+} \frac{x^2 - 7}{x - 7} = \infty$$

9. The correct answer is from the left $-\infty$, from the right ∞.

$$\lim_{x \to \frac{1}{2}} \frac{-3x^2 + 3}{2x - 1} = \frac{-3\left(\frac{1}{2}\right)^2 + 3}{2\left(\frac{1}{2}\right) - 1}$$

$$= \frac{\left(\frac{9}{4}\right)}{0}$$

Since the denominator equals zero and the numerator does not equal zero, there is an asymptote at $x = \frac{1}{2}$. Graphing the function allows us to determine the one-sided limits.

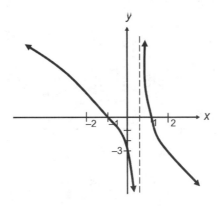

Note: Figure not drawn to scale.

From the left:

$$\lim_{x \to \frac{1}{2}^-} \frac{-3x^2 + 3}{2x - 1} = -\infty$$

From the right:

$$\lim_{x \to \frac{1}{2}^+} \frac{-3x^2 + 3}{2x - 1} = \infty$$

10. The correct answer is from the left ∞, from the right $-\infty$.

$$\lim_{x \to \frac{\sqrt{2}}{2}} \frac{6x - 5}{x^2 - \frac{1}{2}} = \frac{6\left(\frac{\sqrt{2}}{2}\right) - 5}{\left(\frac{\sqrt{2}}{2}\right)^2 - \frac{1}{2}}$$

$$= \frac{3\sqrt{2} - 5}{0}$$

Since the denominator equals zero and the numerator does not equal zero, there is an

asymptote at $x = \frac{\sqrt{2}}{2}$. Graphing the function allows us to determine the one-sided limits.

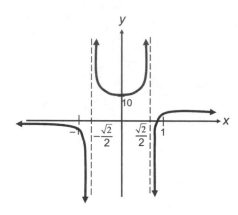

Note: Figure not drawn to scale.

From the left:

$$\lim_{x \to \frac{\sqrt{2}}{2}^-} \frac{6x - 5}{x^2 - \frac{1}{2}} = \infty$$

From the right:

$$\lim_{x \to \frac{\sqrt{2}}{2}^+} \frac{6x - 5}{x^2 - \frac{1}{2}} = -\infty$$

11. The correct answer is $x = 1$.

To find the asymptotes, we are looking for values of x that will result in the denominator equaling zero and the numerator not equaling zero. We can see that when $x = 1$, the function is:

$$g(x) = \frac{4x^3 + 3x^2 + 2x - 1}{x - 1}$$
$$= \frac{4(1)^3 + 3(1)^2 + 2(1) - 1}{(1) - 1}$$
$$= \frac{8}{0}$$

Since the denominator equals zero but the numerator does not equal zero when $x = 1$, there is an asymptote at $x = 1$.

12. The correct answer is none.

To find the asymptotes, we are looking for values of x that will result in the denominator equaling zero and the numerator not equaling zero. To make our job easier, we need to factor the numerator and the denominator:

$$f(x) = \frac{x^3 + x^2 - 9x - 9}{x^2 - 9}$$

$$= \frac{(x + 3)(x - 3)(x + 1)}{(x + 3)(x - 3)}$$

The $x + 3$ and $x - 3$ terms can cancel in the numerator and the denominator, so $x = 3$ and $x = -3$ are not asymptotes. Once these terms have been canceled, we can see that there are no values of x for which the denominator will equal zero and the numerator will not equal zero. So, we conclude that there are no asymptotes.

13. The correct answer is ∞.

    Since this limit consists of a sum of two functions, we examine one function at a time. We can see that when $x = 2$, the first function's denominator equals zero while its numerator does not equal zero. We can conclude that there is a vertical asymptote at $x = 2$. By graphing the first function, we can find its one-sided limit.

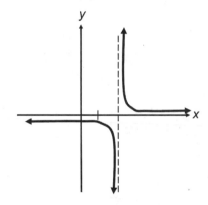

$$\lim_{x \to 2^+} \left(\frac{3}{2x - 4} \right) = \infty$$

The limit of the second function equals a constant:

$$\lim_{x \to 2^+} \left(12x^4 \right) = 12(2)^4$$

$$= 192$$

Using the sum property of infinite limits, we can see:

$$\lim_{x \to 2^+} \left(\frac{3}{2x - 4} + 12x^4 \right) = \infty$$

14. The correct answer is $-\infty$.

Since this limit consists of a product of two functions, we examine one function at a time. We can see that when $x = 1$, the first function's denominator equals zero while its numerator does not equal zero. We can conclude that there is a vertical asymptote at $x = 1$. By graphing the first function, we can find its one-sided limit.

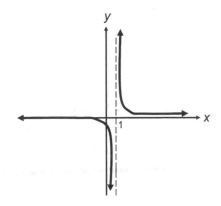

$$\lim_{x \to 1^+} \left(\frac{4}{(x-1)^3} \right) = \infty$$

The limit of the second function equals a negative constant:

$$\lim_{x \to 1^+} \left(-2x^2 - 3 \right) = -2(1)^2 - 3$$
$$= -5$$

Using the product property of infinite limits, we can see:

$$\lim_{x \to 1^+} \left(g(x)h(x) \right) = -\infty$$

15. The correct answer is 0.

Since this limit consists of a quotient of two functions, we examine one function at a time. We can see that when $x = 2$, the function in the numerator equals a constant:

$$\lim_{x \to 2^+} \left(x^2 - 1 \right) = (2)^2 - 1$$
$$= 3$$

We can see that when $x = 2$, the denominator of the second function equals zero while its numerator does not equal zero. We can conclude that there is a vertical asymptote at $x = 2$. By graphing the second function, we can find its one-sided limit.

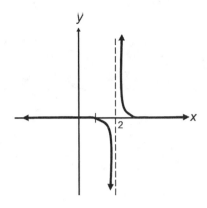

$$\lim_{x \to 2^+} \frac{1}{x-2} = \infty$$

Using the quotient property of infinite limits, we can see:

$$\lim_{x \to 2^+} \frac{x^2 - 1}{\left(\dfrac{1}{x-2}\right)} = \frac{3}{\infty}$$

$$= 0$$

Chapter 6

Differentiation

In this chapter, we'll review the following concepts:

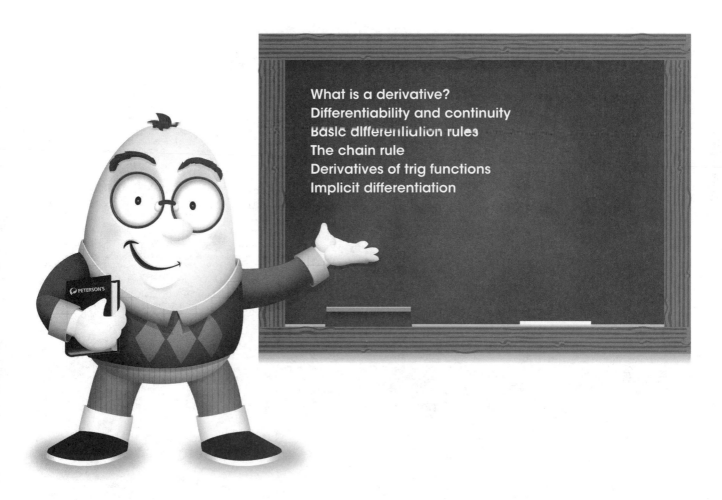

What is a derivative?
Differentiability and continuity
Basic differentiation rules
The chain rule
Derivatives of trig functions
Implicit differentiation

What is a derivative?

The **derivative** of a function is the slope of the function at a particular point. In other words, it is the *rate of change* of the function at that point.

In the figure below, suppose we wanted to find the slope of the curve at the point (x, y_1). We could find the slope by finding the slope of the line that is tangent to the curve at that point:

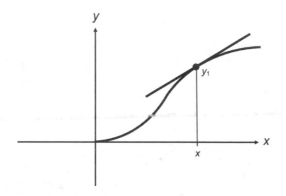

Normally, we would find the slope of a line using two points on the line. But in this case, the tangent line only intersects the graph at one point. We don't have two points to plug into the slope formula. So, we must find the slope another way.

The limit process allows us to find the slope of the tangent line without having two points.

Here's how it works.

egghead's Guide to Calculus

In the figure below, we've added a second point that is *h* units away from *x* on the *x*-axis. The coordinates of the point are $(x + h, y_2)$.

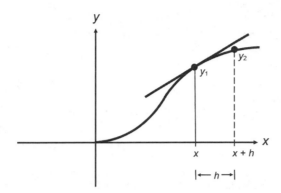

If we were to draw a line between these two points, the slope would be close to the slope of the tangent line to (x, y_1), but not identical. The slope would be a little less steep than the slope of the line we're looking for:

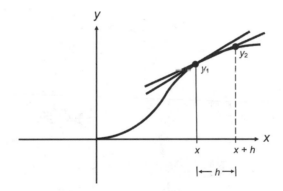

Yet, if we were to move the second point to the left on the curve, bringing it closer to the first point, the slope of the line connecting the points would get closer and closer to the slope of the tangent line. We could determine the slope of the line connecting the points using the slope formula:

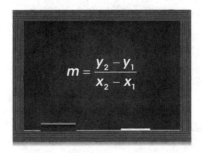

$$m = \frac{y_2 - y_1}{x_2 - x_1}$$

Substituting in the names for the points shown on the graph, we could write the slope formula as follows:

$$m = \frac{y_2 - y_1}{(x + h) - (x)}$$

This just leaves the value of h in the denominator:

$$m = \frac{y_2 - y_1}{h}$$

We could also rewrite the y-coordinates in function notation. The coordinate y_2 would be written as $f(x + h)$. The coordinate y_1 would be written as $f(x)$. This would give us the following formula:

$$m = \frac{f(x + h) - f(x)}{h}$$

This is just a fancy slope formula.

As the points y_2 and y_1 move closer together, the value of h approaches zero. When h equals zero, however, we can no longer use the slope formula, because this would result in division by zero, which is undefined. We get around this problem by using the limit to find the value of the slope:

$$\lim_{h \to 0} \frac{f(x + h) - f(x)}{h}$$

The limit of the function as h approaches zero is the slope of the tangent line to the curve. We call this slope the derivative.

The derivative is noted using any of the following symbols: $f'(x)$, y', or $\frac{dy}{dx}$.

These all mean the same thing: "the derivative of." The symbol $f'(x)$ is pronounced "f prime of x." The symbol y' is pronounced "y prime."

The process of taking the derivative of a function is called **differentiation.**

The symbol $\frac{dy}{dx}$ is pronounced "dee-why-dee-ex."

Differentiability and continuity

If a function is differentiable at a certain point, the function must be continuous at that point. A **continuous** function is one that is not interrupted by gaps or breaks.

Here are some examples of continuous functions:

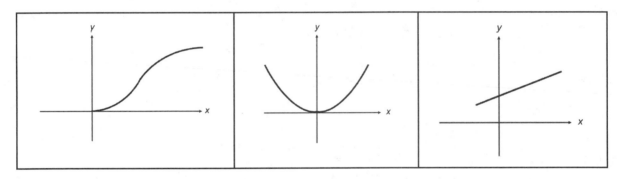

Below are examples of functions that are discontinuous.

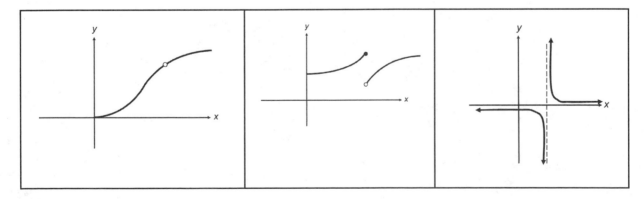

These last three functions are discontinuous because they contain gaps or breaks. They are not differentiable at the points of the gaps or breaks.

Basic differentiation rules

Now that we've reviewed the concept of derivatives, let's look at how to calculate them.

It turns out that there are shortcuts for finding the derivatives of functions. Instead of having to take the limit each time, we can use the following rules.

The **Constant Rule** tells us that for a horizontal line, the derivative is zero.

The line $f(x) = 5$ has no slope.

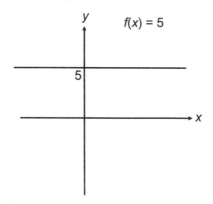

The derivative of a constant is therefore zero.

Irrational numbers such as π, e, and $\sqrt{2}$ are constants.

The **Power Rule** tells us how to differentiate functions with exponents. Take the exponent, move it in front of the variable, and reduce the exponent by 1:

$$f(x) = x^2$$
$$f'(x) = 2x^{2-1}$$
$$= 2x$$

The **Constant Multiple Rule** applies to expressions with constants. Take the exponent, move it in front of the variable, and reduce the exponent by 1. Then multiply by the constant:

$$f(x) = 4x^3$$
$$f'(x) = 4(3x^{3-1})$$
$$= 12x^2$$

Alternatively, the Constant Multiple Rule can be written as follows:

$$f(x) = 4x^3$$
$$f'(x) = (3)4x^{3-1}$$
$$= 12x^2$$

The **Sum and Difference Rules** apply to expressions with added or subtracted terms. Take the derivative of each term individually, and then add or subtract:

$$f(x) = 3x^5 + 7x^4 - 3x^3$$
$$f'(x) = (5)3x^{5-1} + (4)7x^{4-1} - (3)3x^{3-1}$$
$$= 15x^4 + 28x^3 - 9x^2$$

The Product and Quotient Rules are more complex. The **Product Rule** applies to derivatives of multiplied terms. It does not involve taking the derivative of each term separately and then multiplying them. Instead, the formula is as follows:

$$f(x) = (\text{1st term})(\text{2nd term})$$
$$f'(x) = (\text{1st}) * (\text{Derivative of 2nd}) + (\text{2nd}) * (\text{Derivative of 1st})$$

Here is the Product Rule expressed in variables:

$$f(x) = (a)(b)$$
$$f'(x) = (a)(b') + (b)(a')$$

An example would be as follows:

$$f(x) = (7x^3)(3x)$$
$$f'(x) = (7x^3)\left((1)3x^{1-0}\right) + (3x)\left((3)7x^{3-1}\right)$$
$$= (7x^3)(3) + (3x)\left(21x^2\right)$$
$$= 21x^3 + 63x^3$$
$$= 84x^3$$

The **Quotient Rule** applies to derivatives of divided terms. It does not involve taking the derivative of each term separately and then dividing them. Instead, here is the formula:

$$f(x) = \frac{\text{Top term}}{\text{Bottom term}}$$
$$f'(x) = \frac{(\text{Bottom}) * (\text{Derivative of Top}) - (\text{Top}) * (\text{Derivative of Bottom})}{\text{Bottom}^2}$$

Here is the Quotient Rule expressed in variables:

$$f(x) = \frac{a}{b}$$
$$f'(x) = \frac{(b)(a') - (a)(b')}{b^2}$$

Here is an example:

$$f(x) = \frac{2x^2}{3y^3}$$
$$f'(x) = \frac{(3y^3)\left((2)2x^{2-1}\right) - (2x^2)\left((3)3y^{3-1}\right)}{\left(3y^3\right)^2}$$
$$= \frac{(3y^3)(4x) - (2x^2)\left(9y^2\right)}{\left(9y^6\right)}$$
$$= \frac{12y^3x - 18x^2y^2}{9y^6}$$
$$= \frac{4y^3x - 6x^2y^2}{3y^6}$$
$$= \frac{2xy^2(2y - 3x)}{3y^6}$$
$$= \frac{2x(2y - 3x)}{3y^4}$$

Practice Questions—Basic differentiation rules

Directions: Find the derivative of each function. You will find the Practice Question Solutions on page 144.

1. $f(x) = 270e^2$

2. $f(x) = 18x^5$

3. $y = -\dfrac{1}{2}x^8$

4. $g(x) = 5x^4 - \sqrt{\pi}x^2 + 2^e x - 1$

5. $f(x) = \left(x - 4x^4\right)\left(2x^7 - 5\right) + \dfrac{2x^4}{x^2 + 8x - 3}$

The chain rule

The **Chain Rule** is used to differentiate composite functions. Let's say we have two functions, $f(x)$ and $g(x)$. They produce a composite function, $f(g(x))$. The derivative of the composite function is as shown:

$$f'(g(x)) * g'(x)$$

We take the derivative of the outside function first. Then, we multiply that by the derivative of the inside function. We'll see an example of this when we get to trig functions, below.

For functions raised to a power, we use the following formula:

$$f(x) = (u)^c$$
$$f'(x) = c(u)^{c-1}\left(\frac{du}{dx}\right)$$

The quantity $\left(\dfrac{du}{dx}\right)$ is the derivative of u, the quantity in parentheses. We first differentiate $(u)^c$. Then multiply the result by the derivative of u.

Here is an example.

$$f(x) = (2x + 1)^7$$
$$f'(x) = c(u)^{c-1}\left(\frac{du}{dx}\right)$$
$$= (7)(2x + 1)^{7-1}(2)$$
$$= 14(2x + 1)^6$$

This form of the Chain Rule is also called the **General Power Rule.**

Practice Questions—The chain rule

Directions: Find the derivative of each function. You will find the Practice Question Solutions on page 145.

6. $f(x) = (2x^9 + 4x - 4)^2$

7. $f(x) = \dfrac{\left(4x^5 + 3\right)^3}{\left(x^9 - 1\right)^\pi}$

8. $g(x) = \dfrac{\sqrt{x^2 - 7}}{x^3}$

9. $f(x) = \left(\left(3x + 2\right)^2 + 5\right)^2$

10. $h(x) = \left(\left(\left(x - 1\right)^2 + 1\right)^2 + 1\right)^2$

Derivatives of trig functions

The derivatives of trig functions are shown below. The notation $\frac{d}{dx}$ indicates an operation, like the square root sign, $\sqrt{\ }$. When we see the square root sign over a number, such as $\sqrt{2}$, this tells us to take the square root of 2. Similarly, the $\frac{d}{dx}$ notation means "take the derivative of, with respect to x."

The derivatives of the three main trig functions are as follows:

Sine function	$\frac{d}{dx}\sin(x) = \cos(x)$
Cosine function	$\frac{d}{dx}\cos(x) = -\sin(x)$
Tangent function	$\frac{d}{dx}\tan(x) = \sec^2(x)$

The phrase $\frac{d}{dx}\sin(x)$ means take the derivative of $\sin(x)$.

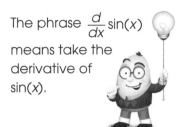

It's a good idea to memorize these three.

The derivatives of the reciprocal trig functions are as follows:

Cosecant function $\left(\frac{1}{\sin(x)}\right)$	$\frac{d}{dx}\csc(x) = -\csc(x)\cot(x)$
Secant function $\left(\frac{1}{\cos(x)}\right)$	$\frac{d}{dx}\sec(x) = \sec(x)\tan(x)$
Cotangent function $\left(\frac{1}{\tan(x)}\right)$	$\frac{d}{dx}\cot(x) = -\csc^2(x)$

For reference, there are a few trig identities that may come in handy when differentiating trig functions. The **Pythagorean Identities** are as follows:

$\sin^2(x) + \cos^2(x) = 1$	$1 + \tan^2(x) = \sec^2(x)$	$1 + \cot^2(x) = \csc^2(x)$

Here are the **Quotient Identities:**

$\tan(x) = \dfrac{\sin(x)}{\cos(x)}$	$\cot(x) = \dfrac{\cos(x)}{\sin(x)}$

Example

To differentiate a trig function, recall the derivatives from memory or look them up on the chart. Take, for example, the following function:

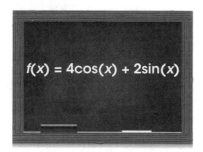

$$f(x) = 4\cos(x) + 2\sin(x)$$

This trig function contains two functions added together. The derivative of $\cos(x)$ is $-\sin(x)$. Multiplying this by the constant, 4, we get $-4\sin(x)$ for the first term. The derivative of $\sin(x)$ is $\cos(x)$. Multiplying this by the constant, 2, we get $2\cos(x)$ for the second term:

$$f(x) = 4\cos(x) + 2\sin(x)$$
$$f'(x) = -4\sin(x) + 2\cos(x)$$

The chain rule and trig functions

The Chain Rule can be used for differentiating composite trig functions. Let's look at the following example:

$$y = \cos\big(\sin(x)\big)$$

To differentiate this function, we start with the function on the outside.

$$\frac{d}{dx}\cos\big(\sin(x)\big) = -\sin\big(\sin(x)\big)$$

This is step one.

Next, take the derivative of the inside function, step two:

$$\frac{d}{dx}\sin(x) = \cos(x)$$

Finally, multiply the results:

$$y = \cos\big(\sin(x)\big)$$
$$y' = -\sin\big(\sin(x)\big) * \cos(x)$$

The Chain Rule can also be used to solve trig functions raised to an exponent.

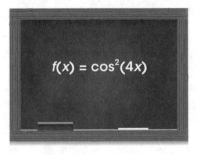

$$f(x) = \cos^2(4x)$$

First, we rewrite the function as follows:

$$f(x) = \big[\cos(4x)\big]^2$$

Next, we apply the General Power Rule. Start with the exponential operation:

$$\frac{d}{dx}[\cos(4x)]^2 = (2)[\cos(4x)]^{2-1}$$
$$= 2[\cos(4x)]^1$$
$$= 2\cos(4x)$$

Then, take the derivative of the expression in brackets, $\cos(4x)$:

$$\frac{d}{dx}\cos(4x) = -\sin(4x)$$

Then, take the derivative of the expression in parentheses, $4x$:

$$\frac{d}{dx}4x = 4$$

Finally, multiply the results:

$$f(x) = \left[\cos(4x)\right]^2$$
$$f'(x) = 2\cos(4x) * -\sin(4x) * 4$$
$$= -8\cos(4x)\sin(4x)$$

Practice Questions—Derivatives of trig functions

Directions: Find the derivatives of the following trigonometric functions. You will find the Practice Question Solutions on page 148.

11. $y = \sqrt{3}\sin(x) - \frac{1}{2}\cos x$

12. $f(x) = \dfrac{5}{\cos(x)}$

13. $y = \pi\cos(x)\cot(x)$

14. $h(x) = \cos(3x + 2)$

15. $u(x) = \sec\left(\cos(2x)\right)$

Implicit differentiation

The technique of **implicit differentiation** is used for equations when y is not isolated on the left side of the equal sign.

Normally, we are used to differentiating equations in the form shown:

$$y = ax^3 + bx^2 + cx + d$$

The y-variable is on the left side of the equation.

The x-variables are on the right.

But sometimes we are given equations in different forms, such as the one shown here:

$$(x + 2)^2 + (y + 1)^2 = 2$$

Here, the y-variable is not isolated on the left. To solve equations like this one, we use implicit differentiation. Here are the steps:

First, differentiate both sides of the equals sign. Use the Chain Rule, and add in multiplying by dx or dy (or whatever variable is in the equation).

$$(x + 2)^2 + (y + 1)^2 = 2$$
$$(2)(x + 2)^{2-1}(1)dx + (2)(y + 1)^{2-1}(1)dy = 0$$
$$(2)(x + 2)dx + (2)(y + 1)dy = 0$$
$$(2x + 4)dx + (2y + 2)dy = 0$$

Remember that the derivative of a constant is zero.

Next, collect all of the dys on one side and all of the dxs on the other:

$$(2x + 4)dx + (2y + 2)dy = 0$$
$$(2y + 2)dy = -(2x + 4)dx$$

Factor out the *dy*s and *dx*s if necessary. In this equation, they are already factored out.

Finally, solve for $\frac{dy}{dx}$:

$$(2y + 2)dy = -(2x + 4)dx$$
$$\frac{dy}{dx} = -\frac{(2x + 4)}{(2y + 2)}$$

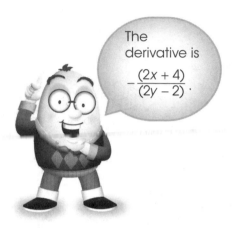

The derivative is $-\frac{(2x + 4)}{(2y - 2)}$.

Some problems might require us to go one step beyond differentiation. We might be asked to find the slope of a tangent line to a function at a particular point. For instance, suppose we were asked to find the slope of the tangent line to the function $(x + 2)^2 + (y + 1)^2 = 2$ at the point $(2, 2)$.

In this case, first we would solve for $\frac{dy}{dx}$ using implicit differentiation, as above. Then we could solve for the slope, which is the value of $\frac{dy}{dx}$ at the requested point. Substitute the coordinates of the point into the equation:

$$\frac{dy}{dx} = -\frac{(2(2) + 4)}{(2(2) - 2)}$$
$$= -\frac{(4 + 4)}{(4 - 2)}$$
$$= -\frac{8}{2}$$
$$= -4$$

The slope at the point $(2, 2)$ is -4.

Finding second derivatives

Certain problems may ask us to find the derivative of a function more than once. These derivatives are known as **higher order derivatives**.

To find the second derivative, we would take the derivative of the function twice. To find the third derivative, we would take the derivative three times, and so on.

As an example, let's find the second derivative of the equation shown.

The second derivative is written $\frac{d^2y}{dx^2}$.

$$4x^2 + 3y - 7x = y + 9$$

First, we solve for $\frac{dy}{dx}$ using implicit differentiation:

$$4x^2 + 3y - 7x = y + 9$$
$$4(2)x^{2-1}dx + 3(1)y^{1-1}dy - 7(1)x^{1-1}dx = (1)y^{1-1}dy + 0$$
$$8xdx + 3dy - 7dx = dy$$
$$3dy - dy = 7dx - 8xdx$$
$$2dy = dx(7 - 8x)$$
$$\frac{dy}{dx} = \frac{(7 - 8x)}{2}$$
$$= \frac{7}{2} - 4x$$

Now we can take the derivative again, but with respect to x, to get the second derivative, $\frac{d^2y}{dx^2}$:

$$\frac{d^2y}{dx^2} = 0 - 4(1)x^{1-1}$$
$$= -4$$

The second derivative is –4.

Practice Questions—Implicit differentiation

Directions: Follow the directions given to solve the problems below. You will find the Practice Question Solutions on page 150.

Use implicit differentiation to find $\frac{dy}{dx}$.

16. $2y + \frac{1}{2}y^3 - x^3 + 2x^2 + y^2 = 1 + x^2 - y^3$

17. $\sqrt{x} + 3x^3 - y + 5 = -\pi^2 x + 5x^2 - y^3$

Find the slope of the tangent line to the graph at the indicated point.

18. $(x + 2)^2 + (y + 1)^2 = 2$, at point $(-3, 0)$

Find the equation of the tangent line to the graph at the indicated point.

19. $y^2 + 2\sin(x) + 1 = \cos(y)$, at point $(5, 1.15)$

Find the second derivative.

20. $2y - 6x^2 + 3x = 12 + y$

Chapter Review

Directions: Use your knowledge of differentiation to solve the problems below. Solutions can be found on page 154.

Find the derivative of each function.

1. $g(x) = 67\pi$

2. $f(x) = 25x^{\frac{7}{5}}$

3. $y = 12x^9 - e^{\frac{3}{2}}x^7 - 8x^2 + \sqrt{7}x + 31$

4. $f(x) = \dfrac{(3x^3 + 3)(x - 1)}{x^3 + x}$

5. $g(x) = (2x^3 + 4x^5 - 6x^7)^2$

6. $h(x) = \dfrac{(x^3 + 3)^3}{3x}$

7. $y = ((2x + 2)^2 + 2)^2$

Find the derivatives of the following trigonometric functions.

8. $j(x) = 83\cos(x) - \sqrt{7}\sin(x) + 3\cos(\pi)$

9. $y = 3\csc(x)\cot(x)$

10. $w(x) = \cot^2(3x)$

11. $f(x) = 38\sin\left(\sec\left(\dfrac{x}{2}\right)\right)$

Use implicit differentiation to find $\dfrac{dy}{dx}$.

12. $ex + 5y - 7y^3 - x^2 = x^7 + \dfrac{5}{2}y - 6$

Find the slope of the tangent line to the graph at the indicated point.

13. $\dfrac{x^2}{9} + y^2 = 2$, at point $\left(2, \dfrac{\sqrt{14}}{3}\right)$

Find the equation of the line that is tangent to the graph at the indicated point.

14. $\sec(x) + \tan(y) + 1 = x$, at point $\left(\dfrac{\pi}{4}, \pi\right)$

Find the second derivative.

15. $x + 2y - 3x^2 = x + 5$

Practice Question Solutions

Basic differentiation rules

1. The correct answer is 0.

We need to recall that e is an irrational constant number. Using the Constant Power Rule, we can see that:

$$f'(x) = 0$$

2. The correct answer is $90x^4$.

Using the Power Rule, we can find the derivative:

$$f'(x) = (5)(18)x^{5-1}$$
$$= 90x^4$$

3. The correct answer is $-4x^7$.

Using the Constant Multiple Rule and the Power Rule, we can find the derivative:

$$\frac{dy}{dx} = \left(-\frac{1}{2}\right)\left(8x^{8-1}\right)$$
$$= \left(-\frac{1}{2}\right)\left(8x^7\right)$$
$$= -4x^7$$

4. The correct answer is $20x^3 - 2\sqrt{\pi}x + 2^e$.

Using a combination of the Sum and Difference Rules, the Constant Multiple Rule, the Power Rule, and the Constant Rule, we can find the derivative:

$$g'(x) = (4)5x^{4-1} - (2)\sqrt{\pi}x^{2-1} + (1)2^e x^{1-1} - 0$$
$$= 20x^3 - 2\sqrt{\pi}x + 2^e$$

5. The correct answer is $-88x^{10} + 16x^7 + 80x^3 - 5 + \dfrac{4x^5 + 48x^4 - 24x^3}{x^4 + 16x^3 + 58x^2 - 48x + 9}$.

Using a combination of all of the basic differentiation rules, including the Product Rule and the Quotient Rule, we can find the derivative:

$$f'(x) = (x - 4x^4)\big((7)2x^{7-1} - 0\big) + (2x^7 - 5)\big(1 - (4)4x^{4-1}\big)$$
$$+ \frac{(x^2 + 8x - 3)\big((4)2x^{4-1}\big) - (2x^4)\big((2)x^{2-1} + (1)8x^{1-1} - 0\big)}{(x^2 + 8x - 3)^2}$$
$$= (x - 4x^4)(14x^6) + (2x^7 - 5)(1 - 16x^3)$$
$$+ \frac{(x^2 + 8x - 3)(8x^3) - (2x^4)(2x + 8)}{(x^2 + 8x - 3)^2}$$
$$= 14x^7 - 56x^{10} + 2x^7 - 32x^{10} - 5 + 80x^3$$
$$+ \frac{8x^5 + 64x^4 - 24x^3 - 4x^5 - 16x^4}{x^4 + 8x^3 - 3x^2 + 8x^3 + 64x^2 - 24x - 3x^2 - 24x + 9}$$
$$= -88x^{10} + 16x^7 + 80x^3 - 5 + \frac{4x^5 + 48x^4 - 24x^3}{x^4 + 16x^3 + 58x^2 - 48x + 9}$$

The chain rule

6. The correct answer is $72x^{17} + 160x^9 - 144x^8 + 32x - 32$.

We can apply the Chain Rule and the other basic differentiation rules to find the derivative:

$$f'(x) = (2)(2x^9 + 4x - 4)^{2-1}\big((9)2x^{9-1} + (1)4x^{1-1} - 0\big)$$
$$= (2)(2x^9 + 4x - 4)(18x^8 + 4)$$
$$= (4x^9 + 8x - 8)(18x^8 + 4)$$
$$= 72x^{17} + 16x^9 + 144x^9 + 32x - 144x^8 - 32$$
$$= 72x^{17} + 160x^9 - 144x^8 + 32x - 32$$

7. The correct answer is $\dfrac{\left(60x^4\right)\left(4x^5+3\right)^2\left(x^9-1\right)^\pi-\left(9\pi x^8\right)\left(4x^5+3\right)^3\left(x^9-1\right)^{\pi-1}}{\left(x^9-1\right)^{2\pi}}$.

Using the Quotient Rule and the Chain Rule, we can find the derivative:

$$f'(x)=\frac{\left(x^9-1\right)^\pi(3)\left(4x^5+3\right)^{3-1}\left((5)4x^{5-1}+0\right)-\left(4x^5+3\right)^3(\pi)\left(x^9-1\right)^{\pi-1}\left((9)x^{9-1}+0\right)}{\left(\left(x^9-1\right)^\pi\right)^2}$$

$$=\frac{\left(x^9-1\right)^\pi(3)\left(4x^5+3\right)^2\left(20x^4\right)-\left(4x^5+3\right)^3(\pi)\left(x^9-1\right)^{\pi-1}\left(9x^8\right)}{\left(x^9-1\right)^{2\pi}}$$

$$=\frac{\left(60x^4\right)\left(4x^5+3\right)^2\left(x^9-1\right)^\pi-\left(9\pi x^8\right)\left(4x^5+3\right)^3\left(x^9-1\right)^{\pi-1}}{\left(x^9-1\right)^{2\pi}}$$

8. The correct answer is $\dfrac{-2x^2+21}{x^6-7x^4}\sqrt{x^2-7}$.

First, we rewrite the function as shown:

$$g(x)=\frac{\sqrt{x^2-7}}{x^3}$$

$$=\frac{\left(x^2-7\right)^{\frac{1}{2}}}{x^3}$$

Now we apply the Quotient Rule and the Chain Rule to find the derivative:

$$g'(x) = \frac{x^3 \left(\frac{1}{2}\right)(x^2-7)^{\frac{1}{2}-1}\left((2)x^{2-1}-0\right)-\left(x^2-7\right)^{\frac{1}{2}}\left((3)x^{3-1}\right)}{\left(x^3\right)^2}$$

$$= \frac{\frac{1}{2}x^3\left(x^2-7\right)^{-\frac{1}{2}}(2x)-\left(x^2-7\right)^{\frac{1}{2}}\left(3x^2\right)}{x^6}$$

$$= \frac{\dfrac{x^4}{\sqrt{x^2-7}}-\left(3x^2\right)\sqrt{x^2-7}}{x^6}$$

$$= \frac{\dfrac{x^4}{\sqrt{x^2-7}}\dfrac{\sqrt{x^2-7}}{\sqrt{x^2-7}}-\left(3x^2\right)\sqrt{x^2-7}}{x^6}$$

$$= \frac{\dfrac{x^4}{x^2-7}\sqrt{x^2-7}-\left(3x^2\right)\sqrt{x^2-7}}{x^6}$$

$$= \frac{\dfrac{x^4}{x^2-7}-3x^2}{x^6}\sqrt{x^2-7}$$

$$= \frac{x^4-3x^2\left(x^2-7\right)}{x^6\left(x^2-7\right)}\sqrt{x^2-7}$$

$$= \frac{x^4-3x^4+21x^2}{x^8-7x^6}\sqrt{x^2-7}$$

$$= \frac{-2x^4+21x^2}{x^8-7x^6}\sqrt{x^2-7}$$

$$= \frac{-2x^2+21}{x^6-7x^4}\sqrt{x^2-7}$$

9. The correct answer is $324x^3 + 648x^2 + 612x + 216$.

We apply the Chain Rule multiple times to find the derivative:

$$f'(x) = 2\left((3x+2)^2+5\right)^{2-1}\left(2(3x+2)^{2-1}\left(3x^{1-1}+0\right)+0\right)$$

$$= \left(2(3x+2)^2+10\right)\left((6x+4)(3)\right)$$

$$= \left(2\left(9x^2+12x+4\right)+10\right)(18x+12)$$

$$= \left(18x^2+24x+18\right)(18x+12)$$

$$= 324x^3+216x^2+432x^2+288x+324x+216$$

$$= 324x^3+648x^2+612x+216$$

10. The correct answer is $8x^7 - 56x^6 + 192x^5 - 400x^4 + 552x^3 - 504x^2 + 288x - 80$.

We apply the Chain Rule multiple times to find the derivative:

$$h'(x) = 2\left(\left((x-1)^2 + 1\right)^2 + 1\right)^{2-1}\left(2\left((x-1)^2 + 1\right)^{2-1}\left(2(x-1)^{2-1}(1-0) + 0\right) + 0\right)$$

$$= 2\left(\left((x-1)^2 + 1\right)^2 + 1\right)\left(2\left((x-1)^2 + 1\right)\left(2(x-1)\right)\right)$$

$$= 2\left(\left(x^2 - 2x + 2\right)^2 + 1\right)\left(2\left(x^2 - 2x + 1 + 1\right)(2x - 2)\right)$$

$$= 2\left(x^4 - 2x^3 + 2x^2 - 2x^3 + 4x^2 - 4x + 2x^2 - 4x + 4 + 1\right)\left(4x^3 - 4x^2 - 8x^2 + 8x + 8x - 8\right)$$

$$= \left(2x^4 - 8x^3 + 16x^2 - 16x + 10\right)\left(4x^3 - 12x^2 + 16x - 8\right)$$

$$= 8x^7 - 24x^6 + 32x^5 - 16x^4 - 32x^6 + 96x^5 - 128x^4 + 64x^3 + 64x^5 - 192x^4 + 256x^3$$
$$-128x^2 - 64x^4 + 192x^3 - 256x^2 + 128x + 40x^3 - 120x^2 + 160x - 80$$
$$= 8x^7 - 56x^6 + 192x^5 - 400x^4 + 552x^3 - 504x^2 + 288x - 80$$

Derivatives of trig functions

11. The correct answer is $\sqrt{3}\cos(x) + \frac{1}{2}\sin x$.

Using the definitions of the derivatives of the sine and cosine functions, as well as the other basic differentiation rules, we can find the derivative:

$$y' = \sqrt{3}\cos(x) + \frac{1}{2}\sin x$$

12. The correct answer is $5\sec(x)\tan(x)$.

We can rewrite the function as:

$$f(x) = \frac{5}{\cos(x)}$$
$$= 5\sec(x)$$

Now we can use the definition of the derivative of the secant function to find the derivative:

$$f'(x) = 5\sec(x)\tan(x)$$

13. The correct answer is $-\pi\cos(x)\left[\csc^2(x) + 1\right]$.

    Here we use the Product Rule and the definitions of the derivatives of the cosine and cotangent functions to find the derivative:

    $$y' = \pi\cos(x)\left(-\csc^2(x)\right) + \cot(x)\left(-\pi\sin(x)\right)$$
    $$= -\pi\cos(x)\csc^2(x) - \pi\sin(x)\frac{\cos(x)}{\sin(x)}$$
    $$= -\pi\cos(x)\csc^2(x) - \pi\cos(x)$$
    $$= -\pi\cos(x)\left[\csc^2(x) + 1\right]$$

14.

The correct answer is $-3\sin(3x + 2)$.

    We can use the Chain Rule and the definition of the derivative of the cosine function to find the derivative:

    $$h'(x) = -\left[\sin(3x + 2)\right](3)$$
    $$= -3\sin(3x + 2)$$

15. The correct answer is $-2\sin(2x)\sec(\cos(2x))\tan(\cos(2x))$.

    We can use the Chain Rule and the definitions of the derivatives of the secant and cosine functions to find the derivative:

    $$u'(x) = \left[\sec(\cos(2x))\tan(\cos(2x))\right](-\sin(2x))(2)$$
    $$= -2\sin(2x)\sec(\cos(2x))\tan(\cos(2x))$$

Implicit differentiation

16. The correct answer is $\dfrac{\left(3x^2 - 2x\right)}{\left(\frac{9}{2}y^2 + 2y + 2\right)}$.

We first take the derivative of each side:

$$2(1)y^{1-1}dy + \frac{1}{2}(3)y^{3-1}dy - (3)x^{3-1}dx + 2(2)x^{2-1}dx + (2)y^{2-1}dy = 0 + (2)x^{2-1}dx - (3)y^{3-1}dy$$

$$2dy + \frac{3}{2}y^2dy - 3x^2dx + 4xdx + 2ydy = 2xdx - 3y^2dy$$

Now we collect all of the *dy*s on one side and all of the *dx*s on the other, and factor them out:

$$2dy + \frac{3}{2}y^2dy + 2ydy + 3y^2dy = 2xdx + 3x^2dx - 4xdx$$

$$dy\left(2 + \frac{3}{2}y^2 + 2y + 3y^2\right) = dx\left(2x + 3x^2 - 4x\right)$$

Finally, we can solve for $\dfrac{dy}{dx}$:

$$\frac{dy}{dx} = \frac{\left(2x + 3x^2 - 4x\right)}{\left(2 + \frac{3}{2}y^2 + 2y + 3y^2\right)}$$

$$= \frac{\left(3x^2 - 2x\right)}{\left(\frac{9}{2}y^2 + 2y + 2\right)}$$

17. The correct answer is $\dfrac{-9x^2 + 10x - \frac{\sqrt{x}}{2x} - \pi^2}{3y^2 - 1}$.

We first take the derivative of each side:

$$\left(\frac{1}{2}\right)x^{\frac{1}{2}-1}dx + 3(3)x^{3-1}dx - (1)y^{1-1}dy + 0 = -\pi^2(1)x^{1-1}dx + 5(2)x^{2-1}dx - (3)y^{3-1}dy$$

$$\frac{1}{2}x^{-\frac{1}{2}}dx + 9x^2dx - dy = -\pi^2dx + 10xdx - 3y^2dy$$

Now we collect all of the *dy*s on one side and all of the *dx*s on the other, and factor them out:

$$-dy + 3y^2dy = -\pi^2dx + 10xdx - \frac{1}{2}x^{-\frac{1}{2}}dx - 9x^2dx$$

$$dy\left(3y^2 - 1\right) = dx\left(-\pi^2 + 10x - \frac{1}{2\sqrt{x}} - 9x^2\right)$$

Now we can solve for $\dfrac{dy}{dx}$:

$$\frac{dy}{dx} = \frac{-\pi^2 + 10x - \dfrac{1}{2\sqrt{x}} - 9x^2}{3y^2 - 1}$$

$$= \frac{-9x^2 + 10x - \dfrac{1}{2\sqrt{x}} \dfrac{\sqrt{x}}{\sqrt{x}} - \pi^2}{3y^2 - 1}$$

$$= \frac{-9x^2 + 10x - \dfrac{\sqrt{x}}{2x} - \pi^2}{3y^2 - 1}$$

18.

The correct answer is 1.

First solve for $\dfrac{dy}{dx}$ using implicit differentiation:

$$(2)(x+2)^{2-1}\big((1)\,dx + 0\big) + (2)(y+1)^{2-1}\big((1)\,dy + 0\big) = 0$$
$$2(x+2)\,dx + 2(y+1)\,dy = 0$$
$$2(y+1)\,dy = -2(x+2)\,dx$$
$$\frac{dy}{dx} = \frac{-2(x+2)}{2(y+1)}$$
$$= \frac{-x-2}{y+1}$$

Now that we have $\dfrac{dy}{dx}$, we can solve for the slope, which is the value of $\dfrac{dy}{dx}$ at the requested point

(–3, 0):

$$\frac{dy}{dx} = \frac{-(-3)-2}{0+1}$$
$$= 1$$

19. The correct answer is $y = -0.8588x + 5.444$.

First solve for $\dfrac{dy}{dx}$ using implicit differentiation:

$$(2)y^{2-1}dy + 2\cos(x)dx + 0 = -\sin(y)dy$$
$$2ydy + 2\cos(x)dx = -\sin(y)dy$$
$$2ydy + \sin(y)dy = -2\cos(x)dx$$
$$dy(2y + \sin(y)) = -2\cos(x)dx$$
$$\frac{dy}{dx} = \frac{-2\cos(x)}{2y + \sin(y)}$$

Substituting the given point in, we can find the slope of the tangent line at that point:

$$\frac{dy}{dx} = \frac{-2\cos(5)}{2(1.15) + \sin(1.15)}$$
$$= -0.8588$$

With the slope of the tangent line and a known point on that line (5, 1.15), we can find the equation of the tangent line using the slope-intercept formula, $y = mx + b$:

$$y = mx + b$$
$$1.15 = (-0.8588)(5) + b$$
$$b = 5.444$$
$$y = -0.8588x + 5.444$$

20. The correct answer is 12.

First, we solve for $\frac{dy}{dx}$ using implicit differentiation:

$$2(1)y^{1-1}dy - 6(2)x^{2-1}dx + 3(1)x^{1-1}dx = 0 + (1)y^{1-1}dy$$
$$2dy - 12xdx + 3dx = dy$$
$$2dy - dy = 12xdx - 3dx$$
$$dy = dx(12x - 3)$$
$$\frac{dy}{dx} = 12x - 3$$

Now we can take the derivative again, but with respect to x, to get the second derivative, $\frac{d^2y}{dx^2}$:

$$\frac{d^2y}{dx^2} = 12(1)x^{1-1} - 0$$
$$= 12$$

Chapter Review Solutions

1. The correct answer is 0.

We need to recall that π is an irrational constant number. Using the Constant Power Rule, we can see that:

$$g'(x) = 0$$

2. The correct answer is $35x^{\frac{2}{5}}$.

Using the Constant Multiple Rule and the Power Rule, we can find the derivative:

$$f'(x) = (25)\left(\frac{7}{5}x^{\frac{7}{5}-1}\right)$$

$$= (5)\left(7x^{\frac{2}{5}}\right)$$

$$= 35x^{\frac{2}{5}}$$

3. The correct answer is $108x^8 - 7e^{\frac{3}{2}}x^6 - 16x + \sqrt{7}$.

Using a combination of the Sum and Difference Rules, the Constant Multiple Rule, the Power Rule, and the Constant Rule, we can find the derivative:

$$y' = (9)12x^{9-1} - (7)e^{\frac{3}{2}}x^{7-1} - (2)8x^{2-1} + (1)\sqrt{7}x^{1-1} + 0$$

$$= 108x^8 - 7e^{\frac{3}{2}}x^6 - 16x + \sqrt{7}$$

4. The correct answer is $\dfrac{3x^6 + 9x^4 - 12x^3 + 9x^2 + 3}{x^6 + 2x^4 + x^2}$.

We can use a combination of all of the basic differentiation rules, including the Product Rule and the Quotient Rule, to find the derivative. Because the numerator of the fraction involves a product, we will have to apply the Product Rule as we are applying the Quotient Rule. This can be somewhat confusing, so it is important to keep close watch on the terms as we proceed.

This is easier to do if we think of the Quotient Rule and Product Rule as follows:

Quotient Rule:

$$\frac{(\text{Bottom}) * (\text{Derivative of Top}) - (\text{Top}) * (\text{Derivative of Bottom})}{\text{Bottom}^2}$$

Product Rule:

$$(\text{1st}) * (\text{Derivative of 2nd}) + (\text{2nd}) * (\text{Derivative of 1st})$$

Applying this thought process, our approach for this problem is:

$$\frac{(\text{Bottom}) * \big((\text{1st})(\text{Derivative of 2nd}) + (\text{2nd})(\text{Derivative of 1st})\big) - (\text{Top}) * (\text{Derivative of Bottom})}{\text{Bottom}^2}$$

Using this approach, we can find the derivative:

$$f'(x) = \frac{(x^3 + x)\left[(3x^3 + 3)(1) + (x-1)\big((3)(3x^{3-1}) + 0\big)\right] - (3x^3 + 3)(x-1)\big((3)(x^{3-1}) + (1)(x^{1-1})\big)}{(x^3 + x)^2}$$

$$= \frac{(x^3 + x)\left[(3x^3 + 3) + (x-1)(9x^2)\right] - (3x^3 + 3)(x-1)(3x^2 + 1)}{(x^3 + x)^2}$$

$$= \frac{(x^3 + x)(3x^3 + 3 + 9x^3 - 9x^2) - (3x^4 - 3x^3 + 3x - 3)(3x^2 + 1)}{x^6 + 2x^4 + x^2}$$

$$= \frac{(x^3 + x)(12x^3 - 9x^2 + 3) - (9x^6 - 9x^5 + 9x^3 - 9x^2 + 3x^4 - 3x^0 + 3x - 3)}{x^6 + 2x^4 + x^2}$$

$$= \frac{12x^6 - 9x^5 + 3x^3 + 12x^4 - 9x^3 + 3x - 9x^6 + 9x^5 - 3x^4 - 6x^3 + 9x^2 - 3x + 3}{x^6 + 2x^4 + x^2}$$

$$= \frac{3x^6 + 9x^4 - 12x^3 + 9x^2 + 3}{x^6 + 2x^4 + x^2}$$

5. The correct answer is $504x^{13} - 576x^{11} - 80x^9 + 128x^7 + 24x^5$.

We can apply the Chain Rule and the other basic differentiation rules to find the derivative:

$$g'(x) = (2)(2x^3 + 4x^5 - 6x^7)^{2-1}\big((3)(2x^{3-1}) + (5)(4x^{5-1}) + (7)(-6x^{7-1})\big)$$

$$= 2(2x^3 + 4x^5 - 6x^7)(6x^2 + 20x^4 - 42x^6)$$

$$= (4x^3 + 8x^5 - 12x^7)(6x^2 + 20x^4 - 42x^6)$$

$$= 24x^5 + 80x^7 - 168x^9 + 48x^7 + 160x^9 - 336x^{11} - 72x^9 - 240x^{11} + 504x^{13}$$

$$= 504x^{13} - 576x^{11} - 80x^9 + 128x^7 + 24x^5$$

6. The correct answer is $\frac{8}{3}x^7 + 15x^4 + 18x - \frac{9}{x^2}$.

Using the Quotient Rule and the Chain Rule, we can find the derivative:

$$h'(x) = \frac{(3x)\left((3)\left(x^3+3\right)^{3-1}\left((3)\left(x^{3-1}\right)+0\right)\right)-\left(x^3+3\right)^3\left((1)\left(3x^{1-1}\right)\right)}{(3x)^2}$$

$$= \frac{(3x)\left((3)\left(x^3+3\right)^2\left(3x^2\right)\right)-\left(x^3+3\right)^3(3)}{9x^2}$$

$$= \frac{27x^3\left(x^3+3\right)^2-3\left(x^3+3\right)^3}{9x^2}$$

$$= \frac{27x^3\left(x^6+6x^3+9\right)-3\left(x^3+3\right)\left(x^6+6x^3+9\right)}{9x^2}$$

$$= \frac{27x^9+162x^6+243x^3+\left(-3x^3-9\right)\left(x^6+6x^3+9\right)}{9x^2}$$

$$= \frac{27x^9+162x^6+243x^3-3x^9-18x^6-27x^3-9x^6-54x^3-81}{9x^2}$$

$$= \frac{24x^9+135x^6+162x^3-81}{9x^2}$$

$$= \frac{24}{9}x^7+\frac{135}{9}x^4+\frac{162}{9}x-\frac{81}{9x^2}$$

$$= \frac{8}{3}x^7+15x^4+18x-\frac{9}{x^2}$$

7. The correct answer is $64x^3 + 192x^2 + 224x + 96$.

We apply the Chain Rule multiple times to find the derivative:

$$y' = 2\left(\left(2x+2\right)^2+2\right)^{2-1}\left(2\left(2x+2\right)^{2-1}\left(2x^{1-1}+0\right)+0\right)$$

$$= 2\left(\left(2x+2\right)^2+2\right)\left(2\left(2x+2\right)(2)\right)$$

$$= \left(2\left(2x+2\right)^2+4\right)(8x+8)$$

$$= 2\left(2x+2\right)^2(8x+8)+4(8x+8)$$

$$= \left(4x^2+8x+4\right)(16x+16)+32x+32$$

$$= 64x^3+64x^2+128x^2+128x+64x+64+32x+32$$

$$= 64x^3+192x^2+224x+96$$

8. The correct answer is $-83\sin(x) - \sqrt{7}\cos(x)$.

Using the definitions of the derivatives of the sine and cosine functions, as well as the other basic differentiation rules, we can find the derivative:

$$j'(x) = -83\sin(x) - \sqrt{7}\cos(x) - 0$$
$$= -83\sin(x) - \sqrt{7}\cos(x)$$

9. The correct answer is $-6\csc^3(x) + 3\csc(x)$.

Using the Product Rule and the definitions of the derivative of the cosecant and cotangent functions, we can find the derivative:

$$y' = 3\csc(x)\left(-\csc^2(x)\right) + \cot(x)\left(-3\csc(x)\cot(x)\right)$$
$$= -3\csc^3(x) - 3\csc(x)\cot^2(x)$$

We can achieve a simpler form if we make use of the following trigonometric identity:

$$1 + \cot^2(x) = \csc^2(x)$$

or

$$\cot^2(x) = \csc^2(x) - 1$$

Substituting in for $\cot^2(x)$, we obtain:

$$y' = -3\csc^3(x) - 3\csc(x)\left(\csc^2(x) - 1\right)$$
$$= -3\csc^3(x) - 3\csc^3(x) + 3\csc(x)$$
$$= -6\csc^3(x) + 3\csc(x)$$

10. The correct answer is $-6\cot(3x) - 6\cot^3(3x)$.

We begin by rewriting the function in a more useful equivalent form:

$$w(x) = \cot^2(3x)$$
$$= \left[\cot(3x)\right]^2$$

We can now use the Chain Rule multiple times and the definition of the derivative of the cotangent function to find the derivative:

$$w'(x) = 2\left[\cot(3x)\right]^{2-1}\left(-\csc^2(3x)\right)\left(3x^{1-1}\right)$$
$$= -6\cot(3x)\csc^2(3x)$$

We can achieve a simpler form if we make use of the following trigonometric identity:

$$\csc^2(x) = \cot^2(x) + 1$$

Substituting in for $\csc^2(x)$, we obtain:

$$w'(x) = -6\cot(3x)\left(1 + \cot^2(3x)\right)$$
$$= -6\cot(3x) - 6\cot^3(3x)$$

11. The correct answer is $19\cos\left(\sec\left(\frac{x}{2}\right)\right)\sec\left(\frac{x}{2}\right)\tan\left(\frac{x}{2}\right)$.

We can use the Chain Rule and the definitions of the derivatives of the sine and secant functions to find the derivative:

$$f'(x) = 38\cos\left(\sec\left(\frac{x}{2}\right)\right)\left(\sec\left(\frac{x}{2}\right)\tan\left(\frac{x}{2}\right)\right)\left(\frac{1}{2}x^{1-1}\right)$$
$$= 19\cos\left(\sec\left(\frac{x}{2}\right)\right)\sec\left(\frac{x}{2}\right)\tan\left(\frac{x}{2}\right)$$

12. The correct answer is $\dfrac{-7x^6 - 2x + e}{21y^2 - \frac{5}{2}}$.

We first take the derivative of each side:

$$e\left(x^{1-1}\right)dx + 5\left(y^{1-1}\right)dy - 7(3)\left(y^{3-1}\right)dy - (2)\left(x^{2-1}\right)dx = (7)\left(x^{7-1}\right)dx + \frac{5}{2}\left(y^{1-1}\right)dy + 0$$
$$edx + 5dy - 21y^2dy - 2xdx = 7x^6dx + \frac{5}{2}dy$$

Now we collect all of the *dy*s on one side and all of the *dx*s on the other, and factor them out:

$$edx - 2xdx - 7x^6dx = \frac{5}{2}dy - 5dy + 21y^2dy$$
$$dx\left(e - 2x - 7x^6\right) = dy\left(\frac{5}{2} - 5 + 21y^2\right)$$

Finally, we can solve for $\dfrac{dy}{dx}$:

$$\frac{dy}{dx} = \frac{\left(e - 2x - 7x^6\right)}{\left(\frac{5}{2} - 5 + 21y^2\right)}$$
$$= \frac{-7x^6 - 2x + e}{21y^2 - \frac{5}{2}}$$

13. The correct answer is $-\dfrac{\sqrt{14}}{21}$.

First solve for $\dfrac{dy}{dx}$ using implicit differentiation:

$$\frac{1}{9}(2)x^{2-1}dx + (2)y^{2-1}dy = 0$$

$$\frac{2}{9}xdx + 2ydy = 0$$

$$2ydy = -\frac{2}{9}xdx$$

$$\frac{dy}{dx} = \frac{-\frac{2}{9}x}{2y}$$

$$= -\frac{x}{9y}$$

Now that we have $\dfrac{dy}{dx}$, we can solve for the slope, which is the value of $\dfrac{dy}{dx}$ at the requested

point $\left(2, \dfrac{\sqrt{14}}{3}\right)$:

$$\frac{dy}{dx} = -\frac{(2)}{9\left(\frac{\sqrt{14}}{3}\right)}$$

$$= -\frac{2}{3\sqrt{14}}$$

$$= -\frac{2}{3\sqrt{14}}\frac{\sqrt{14}}{\sqrt{14}}$$

$$= -\frac{2\sqrt{14}}{3(14)}$$

$$= -\frac{\sqrt{14}}{21}$$

14. The correct answer is $y = \left(1-\sqrt{2}\right)x + \dfrac{3\pi + \pi\sqrt{2}}{4}$.

First solve for $\dfrac{dy}{dx}$ using implicit differentiation:

$$\sec(x)\tan(x)dx + \sec^2(y)dy + 0 = dx$$

$$\sec^2(y)dy = dx - \sec(x)\tan(x)dx$$

$$\sec^2(y)dy = (1-\sec(x)\tan(x))dx$$

$$\frac{dy}{dx} = \frac{1-\sec(x)\tan(x)}{\sec^2(y)}$$

Substituting in for the given point, we can find the slope of the tangent line at that point:

$$\frac{dy}{dx} = \frac{1 - \sec\left(\frac{\pi}{4}\right)\tan\left(\frac{\pi}{4}\right)}{\sec^2(\pi)}$$

$$= \frac{1 - \dfrac{1}{\cos\left(\frac{\pi}{4}\right)}\dfrac{\sin\left(\frac{\pi}{4}\right)}{\cos\left(\frac{\pi}{4}\right)}}{\dfrac{1}{\cos^2(\pi)}}$$

$$= \frac{1 - \dfrac{1}{\frac{\sqrt{2}}{2}}\dfrac{\frac{\sqrt{2}}{2}}{\frac{\sqrt{2}}{2}}}{\dfrac{1}{(-1)^2}}$$

$$= 1 - \frac{1}{\frac{\sqrt{2}}{2}}$$

$$= 1 - \frac{2}{\sqrt{2}}$$

$$= 1 - \frac{2}{\sqrt{2}}\frac{\sqrt{2}}{\sqrt{2}}$$

$$= 1 - \frac{2}{2}\sqrt{2}$$

$$= 1 - \sqrt{2}$$

With the slope of the tangent line and a known point on that line $\left(\frac{\pi}{4}, \pi\right)$, we can find the equation of the tangent line:

$$y = mx + b$$

$$\pi = \left(1 - \sqrt{2}\right)\left(\frac{\pi}{4}\right) + b$$

$$\pi = \frac{\pi}{4} - \frac{\pi\sqrt{2}}{4} + b$$

$$b = \pi - \frac{\pi}{4} + \frac{\pi\sqrt{2}}{4}$$

$$= \frac{4\pi}{4} - \frac{\pi}{4} + \frac{\pi\sqrt{2}}{4}$$

$$= \frac{3\pi}{4} + \frac{\pi\sqrt{2}}{4}$$

$$= \frac{3\pi + \pi\sqrt{2}}{4}$$

$$y = \left(1 - \sqrt{2}\right)x + \frac{3\pi + \pi\sqrt{2}}{4}$$

15.

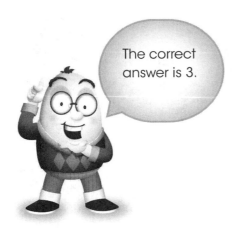

The correct answer is 3.

First, solve for $\dfrac{dy}{dx}$ using implicit differentiation:

$$(1)\,x^{1-1}dx + 2(1)\,y^{1-1}dy - 3(2)\,x^{2-1}dx = (1)\,x^{1-1}dx + 0$$
$$dx + 2dy - 6xdx = dx$$
$$2dy = dx - dx + 6xdx$$
$$2dy = 6xdx$$
$$\frac{dy}{dx} = \frac{6x}{2}$$
$$= 3x$$

Now we can take the derivative again, but with respect to x, to get the second derivative, $\dfrac{d^2y}{dx^2}$:

$$\frac{d^2y}{dx^2} = 3(1)x^{1-1}$$
$$= 3$$

Chapter 7

Applying Differentiation

**In this chapter, we'll review
the following concepts:**

Rates of change
Extrema on an Interval
Increasing and decreasing functions

Rates of change

The technique of differentiation can be used to solve a multitude of problems in the real world. In this chapter, we'll review three types of application problems for derivatives. We'll start with problems involving rates of change.

Rates of change problems typically involve three types of functions:

1. the position function

2. the velocity function

3. the acceleration function

The **position function** of a moving object indicates the distance that object moves with respect to time.

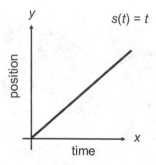

It is frequently denoted by the letter s. So, the position function will typically be written $s(t)$.

To find the **velocity** at which the object is traveling, we take the derivative of the position function.

In other words, we find the slope.

Velocity indicates the rate of change of a position function. It tells us the rate at which position is changing with respect to time. So its value is the first derivative of the position function:

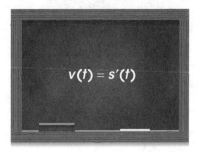

$$v(t) = s'(t)$$

To find the **acceleration** of the object, we take the derivative of the velocity function.

$$a(t) = v'(t)$$

Acceleration is the rate of change of velocity. It tells us how quickly or slowly the velocity is changing.

Acceleration can also be determined by taking the second derivative of the position function. (This is the same as taking the first derivative of velocity.)

$$a(t) = s''(t)$$

To recap, here are the important functions in a summary table:

Function	What It Means	How It's Written	How to Find It
Position	Position (such as distance or height) with respect to time	$s(t)$	given
Velocity	Rate of change of the position function	$v(t)$	$v(t) = s'(t)$
Acceleration	Rate of change of the velocity function	$a(t)$	$a(t) = v'(t)$ or $a(t) = s''(t)$

Rates of change of free-falling objects

If an object is falling, we use the position function for a free-falling object. The known position function for any free-falling object is:

$$s(t) = \frac{1}{2}gt^2 + v_0 t + s_0$$

In this formula, the variables represent the following quantities:

- $s(t)$ is the position of the object.

- g represents acceleration due to gravity.

- t represents time.

- v_0 represents the initial velocity of the object.

- s_0 represents the initial height of the object.

The acceleration due to gravity on Earth is $-32\frac{\text{ft.}}{\text{s}^2}$. Acceleration may change under different gravitational conditions.

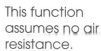

This function assumes no air resistance.

Other types of rates of change

Rates of change problems commonly have to do with objects moving a certain distance as a function of time. However, they can also be defined in terms of other parameters. A rate of change problem might concern the output produced by a machine over time. Or, it might involve the amount of water that flows out of a tank as the number of holes in the tank increases. Any number of (x, y) combinations is possible.

In each case, the position function expresses the relationship between x and y. It describes how y varies in terms of x.

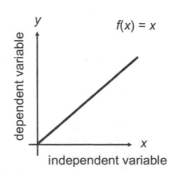

The velocity function expresses the change in y as a function of x.

$$\frac{\text{velocity}}{\text{function}} = \frac{\text{slope of}}{\text{position function}} = \frac{dy}{dx}$$

The acceleration function expresses the rate of change of velocity:

$$\frac{\text{acceleration}}{\text{function}} = \frac{\text{slope of}}{\text{velocity function}} = \frac{d^2y}{dx^2}$$

Example

Here is an example of a problem involving rates of change.

You stand on the edge of a cliff that is 55 feet tall and throw a stone straight up in the air, with the stone leaving your hand when it is 5 feet above the surface of the cliff. The stone just misses the edge of the cliff on its way down and continues to fall until it hits a lake that is at the bottom of the cliff.

The position of the stone in feet can be expressed using the following equation:

$$s(t) = -16t^2 + 16t + 60$$

In this case, the time $t = 0$ is the moment the stone leaves your hand.

Answer the following questions.

(a) What is the initial velocity of the stone?

(b) What is the velocity of the stone at just the moment that it hits the lake?

(c) What is the acceleration of the stone?

To find the initial velocity of the stone, we compare the provided position function to our known position function for a free-falling object.

Our known position function for any free-falling object is:

$$s(t) = \frac{1}{2}gt^2 + v_0 t + s_0$$

And the provided position function is:

$$s(t) = -16t^2 + 16t + 60$$

Comparing the two functions, we see that in our case, the initial velocity is $16\frac{\text{ft.}}{\text{s}}$. It is positive because initially the stone was thrown upwards, and upwards velocities are positive. (Downward velocities are negative.)

To find the velocity of the stone just before it hits the lake, we need to use the position function to obtain the velocity function. This can be achieved by taking the derivative of the position function:

$$s(t) = -16t^2 + 16t + 60$$
$$v(t) = (2)\left(-16t^{2-1}\right) + (1)\left(16t^{1-1}\right) + 0$$
$$= -32t + 16$$

Now we have a function that tells us the velocity of the stone at any time. However, we don't yet know the time at which the stone will hit the lake. To find this, we make use of the position function, substituting in the position at the lake (which is 0) and solving for the time.

$$0 = -16t^2 + 16t + 60$$
$$0 = -16\left(t - \frac{5}{2}\right)\left(t + \frac{3}{2}\right)$$
$$t - \frac{5}{2} = 0, \ t + \frac{3}{2} = 0$$
$$t = \frac{5}{2}, \ t = -\frac{3}{2}$$

We can neglect the negative time answer (because negative time doesn't make any sense), and so we conclude that the stone hits the water at $t = \frac{5}{2}$ seconds. Substituting this into our velocity function, we find the velocity of the stone just before it hits the water:

$$v\left(\frac{5}{2}\right) = -32\left(\frac{5}{2}\right) + 16$$
$$= -80 + 16$$
$$= -64\frac{\text{ft.}}{\text{s}}$$

To find the acceleration of the stone, we can take the derivative of the velocity function:

$$v(t) = -32t + 16$$
$$a(t) = (1)(-32t^{1-1}) + 0$$
$$= -32\frac{\text{ft.}}{\text{s}^2}$$

This makes sense because we are on Earth, and we know that the acceleration due to gravity on Earth is $-32\frac{\text{ft.}}{\text{s}^2}$.

The correct answers are (a) $16\frac{\text{ft.}}{\text{s}}$, (b) $-64\frac{\text{ft.}}{\text{s}}$, and (c) $-32\frac{\text{ft.}}{\text{s}^2}$.

The derivative of velocity is acceleration.

Practice Questions—Rates of change

Directions: Solve each problem for the requested values. You will find the Practice Question Solutions on page 181.

1. You use an air cannon to shoot a ball straight up into the air. It leaves your canon when it is 5 feet off the ground at an initial velocity of $242\frac{\text{ft.}}{\text{s}}$. Answer the following questions:

   (a) How long will it take for the ball rise to 150 feet?

   (b) At what time will the ball have continued to its peak and then descended back to 150 feet?

   (c) What will be its velocity when it is on its way down and passes the 150-foot height?

2. The surface area of a sphere is given by: $SA = 4\pi r^2$, where r is the radius of the sphere. Find the rate of change of the surface area with respect to the radius of the sphere, when the radius of the sphere is 12 meters.

3. The equation that describes the force of gravitational attraction between two planets is $F_G = \frac{Gm_1m_2}{r^2}$, where G is a constant equal to $6.67 \times 10^{-11}\frac{\text{Nm}^2}{\text{kg}^2}$, m_1 and m_2 are the masses of the two planets in kilograms, and r is the distance between the centers of the two planets in meters. Given Jupiter (mass = 1.898×10^{27} kg) and Saturn (mass = 568.3×10^{24} kg), what is the rate of change of the gravitational force between the two planets with respect to distance of separation when they are at their closest to each other (r = 680,000,000,000 m)?

Extrema on an interval

Another type of application problem involving differentiation focuses on finding extrema on an interval.

An **interval** is a section of the graph of a function that lies between two specific endpoints. A **closed interval** includes the two endpoints.

The closed interval shown refers to the portion of a graph that lies between the *x*-value –5 and the *x*-value 4:

The values –5 and 4 are included on the interval.

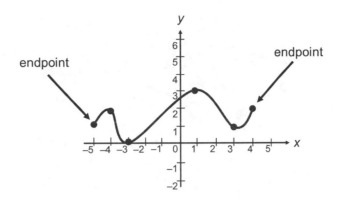

The figure shown above contains the graph of a function with six coordinates marked. The two endpoints are at x-values –5 and 4. The graph also contains high points and low points at x-values –4, –3, 1, and 3.

All of the high and low points are called **extrema**.

The highest and lowest extrema on the interval are called **absolute extrema.** On the graph, the absolute extrema are located at points (–3, 0) and (1, 3).

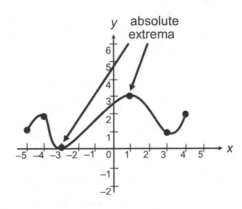

The other extrema on the graph are called **relative extrema** or **local extrema.** These are points that represent high or low points on the interval, but not the highest or lowest points. On the graph, the relative extrema are located at points (–4, 2) and (3, 1).

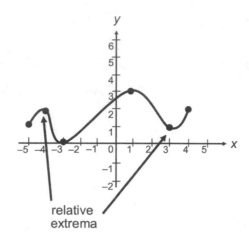

relative
extrema

Absolute or relative high points are called **maxima**, and absolute or relative low points are called **minima**. So, extrema can be either absolute or relative, and they can be either maxima or minima.

Here are the labels for all six points.

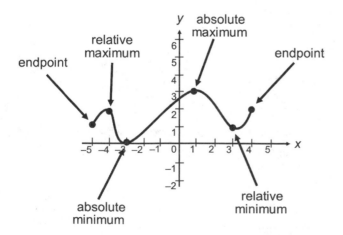

endpoint

relative
maximum

absolute
maximum

endpoint

absolute
minimum

relative
minimum

It is possible to have more than one absolute maximum or minimum on a graph.

Critical numbers

Extrema always occur on a graph at points where the slope of the tangent line to the graph is 0, or at points where the slope does not exist. On the graph shown, all four extrema occur at points where the slope of the tangent line to the graph is a horizontal line.

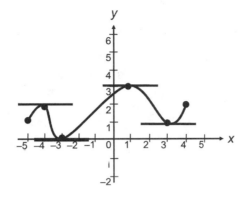

The x-values of extrema are known as **critical numbers.** This graph contains critical numbers –4, –3, 1, and 3.

We use critical numbers to solve extrema problems.

Example

To work through an extrema on an interval problem such as the one below, we would complete the following steps.

> Find the extrema on the indicated closed interval.
> Also indicate which extrema are absolute minimums and maximums.
>
> $y = 3x^3 + 2x^2 - 5x - 3$, within the closed interval [–2, 1]

To find the extrema, we need to find all values of x where the derivative equals zero or where the function is undefined. We start by taking the derivative:

$$y' = (3)(3x^{3-1}) + (2)(2x^{2-1}) - (1)(5x^{1-1}) - 0$$
$$= 9x^2 + 4x - 5$$

Now we set the derivative equal to zero, and solve:

$$0 = 9x^2 + 4x - 5$$
$$0 = (9x - 5)(x + 1)$$
$$9x - 5 = 0, \ x + 1 = 0$$
$$9x = 5, \ x = -1$$
$$x = \frac{5}{9}, \ x = -1$$

We can see from the function that it is defined for all values of x. (It does not involve a fraction where there might be the possibility of division by zero, and there are no square roots where there might be

the possibility of there being a negative square root). So the only critical numbers are $x = \frac{5}{9}$ and $x = -1$. We will find the values of the function at each of these critical points and at the two endpoints of the given region (–2 and 1).

At –2, the function equals:

$$f(-2) = 3(-2)^3 + 2(-2)^2 - 5(-2) - 3$$
$$= -24 + 8 + 10 - 3$$
$$= -9$$

At –1, the function equals:

$$f(-1) = 3(-1)^3 + 2(-1)^2 - 5(-1) - 3$$
$$= -3 + 2 + 5 - 3$$
$$= 1$$

At $\frac{5}{9}$, the function equals:

$$f\left(\frac{5}{9}\right) = 3\left(\frac{5}{9}\right)^3 + 2\left(\frac{5}{9}\right)^2 - 5\left(\frac{5}{9}\right) - 3$$

$$= 3\left(\frac{125}{729}\right) + 2\left(\frac{25}{81}\right) - \frac{25}{9} - 3$$

$$= \frac{125}{243} + \frac{50}{81} - \frac{25}{9} - 3$$

$$= \frac{125}{243} + \frac{150}{243} - \frac{675}{243} - \frac{729}{243}$$

$$= -\frac{1129}{243}$$

$$\approx -4.646$$

Finally, substitute in the value of $x = 1$.

At 1, the function equals:

$$f(1) = 3(1)^3 + 2(1)^2 - 5(1) - 3$$

$$= 3 + 2 - 5 - 3$$

$$= -3$$

Substituting the four values of x gives us four coordinates. We can see that we have an absolute maximum at (–1, 1) and an absolute minimum at (–2 ,–9). The other extrema in this interval are

(1, –3) and $\left(\frac{5}{9}, -\frac{1129}{243}\right)$.

Practice Questions—Extrema on an interval

Directions: Find the extrema on the indicated closed interval. Also indicate which extrema are absolute minimums and maximums. You will find the Practice Question Solutions on page 183.

4. $y = \frac{1}{5}x^5 - \frac{3}{4}x^4 + \frac{1}{3}x^3 + \frac{3}{2}x^2 - 2x + 6$, within the closed interval [–2, 3]

5. $y = \sin(x) + 3$, within the closed interval [0, 2π]

6. $y = -\cos(2x) - \sqrt{2}x$, within the closed interval $\left[0, \frac{\pi}{2}\right]$

Increasing and decreasing functions

A third type of derivative application problem addresses increasing and decreasing functions.

To solve problems involving increasing and decreasing functions, we use the derivative. This process is called **the first derivative test.**

Here are the steps.

1. Take the derivative of the function.

2. Set the derivative equal to zero and solve for the value(s) of x.

3. The values of x are the critical numbers of the function. They separate the function into different intervals. Identify the intervals of the function based on the critical numbers.

egghead's Guide to Calculus

4. For each interval, choose a test point. Substitute that test point into the derivative to determine the slope.

5. If the value is positive, the slope is increasing on that interval. If the value at the test point is negative, the slope is decreasing on that interval.

Once we know whether the graph is increasing or decreasing on an interval, we can determine which x-values are relative maxima or minima of the graph. Then, we can substitute the x-values into the function to find the y-values of the relative maxima or minima.

Example

Consider the function shown:

$$f(x) = 3x^2 + 2$$

For this function, we are to find the intervals where the function is increasing or decreasing.

We can see that there are no values of x for which the function is undefined. We need to find the critical numbers, so we will take the derivative and set it equal to zero:

$$f'(x) = (2)(3x^{2-1}) + 0$$
$$= 6x$$
$$0 = 6x$$
$$x = 0$$

So $x = 0$ is the only critical number. This critical number separates the function into two different intervals, $(-\infty, 0)$ and $(0, \infty)$. To determine if the graph of the function is increasing or decreasing in each interval, we will choose a test point in each region and determine the value of the derivative at that test point.

Region 1, $(-\infty, 0)$:

Test Point: $x = -1$

$$f'(-1) = 6(-1)$$
$$= -6$$

Region 2, $(0, \infty)$:

Test Point: $x = 1$

$$f'(1) = 6(1)$$
$$= 6$$

Since the derivative (and therefore the slope) is negative in the first region, we conclude that the graph is decreasing in that region. Since the derivative is positive in the second region, we conclude that the graph is increasing in that region.

The correct answer is that the function is decreasing for $(-\infty, 0)$.

It Is Increasing for $(0, \infty)$.

Practice Questions—Increasing and decreasing functions

Directions: Solve the problems following the directions indicated. You will find the Practice Question Solutions on page 186.

Find the intervals where the function is increasing or decreasing.

7. $f(x) = x^3 - 5x^2 - \dfrac{1}{x}$

Find the relative extrema of the function on the indicated interval.

8. $y = 7\sin(x) - 3\cos(x + 2)$, on the interval $(0, 2\pi)$

9. $f(x) = 2\tan(x) - \dfrac{1}{2}\tan\left(\dfrac{1}{2}x\right)$, on the interval $(0, \pi)$

Chapter Review

Directions: Solve each problem for the requested values. Solutions can be found on page 189.

1. A liquid helium reservoir on a spacecraft is made up of a combination of a spherical tank, whose volume is given by the formula $V_s = \frac{4}{3}\pi r_s^3$, and a cylindrical accumulator, whose volume is given by $V_c = \pi r_c^2 h$, where the radius of the cylinder is chosen to always equal $\frac{1}{4}$ the radius of the sphere, and the height of the cylinder is fixed at 0.1 meters. Find the rate of change of the volume in the reservoir with respect to the radius of the sphere, when the radius of the sphere equals 0.2 meters.

Find the extrema on the indicated closed interval. Also indicate which extrema are absolute minimums and maximums.

2. $y = x + 2\cos(x)$, $\left[0, \frac{3\pi}{2}\right]$

Find the intervals where the function is increasing or decreasing.

3. $f(x) = x^4 - \sqrt{x}$

Practice Question Solutions

Rates of change

1. The correct answers are (a) 0.625 seconds, (b) 14.5 seconds, and (c) $-222 \frac{\text{ft.}}{\text{s}}$.

   First, we must find a position function for the ball. Our known position function for any free-falling object is:

   $$s(t) = \frac{1}{2} g t^2 + v_0 t + s_0$$

   Substituting in our known values for g (which equals $-32 \frac{\text{ft.}}{\text{s}^2}$), v_0 (which equals $242 \frac{\text{ft.}}{\text{s}}$), and s_0 (which equals 5 ft.), we can find our position function for the ball:

   $$s(t) = \frac{1}{2}(-32)t^2 + (242)t + (5)$$
   $$= -16t^2 + 242t + 5$$

   Now we can substitute 150 feet for the position and solve for the time:

$$150 = -16t^2 + 242t + 5$$
$$0 = -16t^2 + 242t - 145$$
$$0 = (8t - 5)(-2t + 29)$$
$$8t - 5 = 0, \quad -2t + 29 = 0$$
$$8t = 5, \quad -2t = -29$$
$$t = \frac{5}{8}, \quad t = \frac{29}{2}$$
$$t = 0.625 \text{ s}, \quad t = 14.5 \text{ s}$$

The smaller of the two times, 0.625 seconds, is the time it takes for the ball to ascend initially to the 150-foot height. The larger of the two times, 14.5 seconds, is the time it takes for the ball to reach its peak height and then descend back to 150 feet.

In order to find the velocity when the ball is descending back through the 150-foot height, we need to find the velocity function. We do this by taking the derivative of the position function:

$$s(t) = -16t^2 + 242t + 5$$
$$v(t) = (2)(-16t^{2-1}) + (1)(242t^{1-1}) + 0$$
$$= -32t + 242$$

Now we can substitute in $t = 14.5$ seconds, which we found previously is the time when the ball is descending back through 150 feet:

$$v(14.5) = -32(14.5) + 242$$
$$= -222 \frac{\text{ft.}}{\text{s}}$$

2. The correct answer is $96\pi \dfrac{m^2}{m}$.

We are looking for $\dfrac{dSA}{dr}$. To find it, we will take the derivative of the SA function with respect to r:

$$SA = 4\pi r^2$$
$$\frac{dSA}{dr} = (2)\left(4\pi r^{2-1}\right)$$
$$= 8\pi r$$

We can now substitute in $r = 12$ to find the rate of change of the surface area with respect to the radius when the radius equals 12 meters:

$$\frac{dSA}{dr} = 8\pi(12)$$
$$= 96\pi \frac{m^2}{m}$$

We have given the units as $\dfrac{m^2}{m}$, because what we have found is the rate of change of the surface area (m^2) with respect to change in radius (m).

3. The correct answer is $-4.58 \times 10^8 \dfrac{N}{m}$.

We are looking for $\dfrac{dF_G}{dr}$. To find it, we will take the derivative of the F_G function with respect to r:

$$F_G = \frac{Gm_1m_2}{r^2}$$
$$= Gm_1m_2r^{-2}$$
$$\frac{dF_G}{dr} = (-2)Gm_1m_2r^{-2-1}$$
$$= -2Gm_1m_2r^{-3}$$
$$= -\frac{2Gm_1m_2}{r^3}$$

Now we can substitute in the values for the masses of the two planets, the constant G, and the closest distance of separation between the planets to find the desired rate:

$$\frac{dF_G}{dr} = -\frac{2\left(6.67 \times 10^{-11}\right)\left(1.898 \times 10^{27}\right)\left(568.3 \times 10^{24}\right)}{(680,000,000,000)^3}$$
$$= -4.58 \times 10^8 \frac{N}{m}$$

Extrema on an interval

4. The correct answer is that there is an absolute maximum at (3, 10.35), an absolute minimum at (–2, –5.067), and the other extrema in this interval are (–1, 8.217), (1, 5.283), and (2, 5.067).

To find the extrema, we need to find all values of x where the derivative equals zero or where the function is undefined. We start by taking the derivative:

$$y' = (5)\left(\frac{1}{5}x^{5-1}\right) - (4)\left(\frac{3}{4}x^{4-1}\right) + (3)\left(\frac{1}{3}x^{3-1}\right) + (2)\left(\frac{3}{2}x^{2-1}\right) - (1)\left(2x^{1-1}\right) + 0$$

$$= x^4 - 3x^3 + x^2 + 3x - 2$$

Now we set the derivative equal to zero, and solve:

$$0 = x^4 - 3x^3 + x^2 + 3x - 2$$

$$0 = (x \quad 1)(x \quad 1)(x + 1)(x - 2)$$

$$x = 1, \ x = -1, \ x = 2$$

We can see from the function that it is defined for all values of x. So the critical numbers are $x = -1$, $x = 1$, and $x = 2$. We will find the values of the function at each of these critical points and at the two endpoints of the given region (–2 and 3).

At $x = -2$, the function equals:

$$f(-2) = \frac{1}{5}(-2)^5 - \frac{3}{4}(-2)^4 + \frac{1}{3}(-2)^3 + \frac{3}{2}(-2)^2 - 2(-2) + 6$$

$$= -\frac{32}{5} - 12 - \frac{8}{3} + 6 + 4 + 6$$

$$\approx -5.067$$

At $x = -1$, the function equals:

$$f(-1) = \frac{1}{5}(-1)^5 - \frac{3}{4}(-1)^4 + \frac{1}{3}(-1)^3 + \frac{3}{2}(-1)^2 - 2(-1) + 6$$

$$= -\frac{1}{5} - \frac{3}{4} - \frac{1}{3} + \frac{3}{2} + 2 + 6$$

$$\approx 8.217$$

At $x = 1$, the function equals:

$$f(1) = \frac{1}{5}(1)^5 - \frac{3}{4}(1)^4 + \frac{1}{3}(1)^3 + \frac{3}{2}(1)^2 - 2(1) + 6$$

$$= \frac{1}{5} - \frac{3}{4} + \frac{1}{3} + \frac{3}{2} - 2 + 6$$

$$\approx 5.283$$

At $x = 2$, the function equals:

$$f(2) = \frac{1}{5}(2)^5 - \frac{3}{4}(2)^4 + \frac{1}{3}(2)^3 + \frac{3}{2}(2)^2 - 2(2) + 6$$

$$= \frac{32}{5} - 12 + \frac{8}{3} + 6 - 4 + 6$$

$$\approx 5.067$$

At $x = 3$, the function equals:

$$f(3) = \frac{1}{5}(3)^5 - \frac{3}{4}(3)^4 + \frac{1}{3}(3)^3 + \frac{3}{2}(3)^2 - 2(3) + 6$$

$$= \frac{243}{5} - \frac{243}{4} + 9 + \frac{27}{2} - 6 + 6$$

$$= 10.35$$

We can see that we have an absolute maximum at (3, 10.35) and an absolute minimum at (–2, –5.067). The other extrema in this interval are (–1, 8.217), (1, 5.283), and (2, 5.067).

5. The correct answer is an absolute maximum at $\left(\frac{\pi}{2}, 4\right)$, an absolute minimum at $\left(\frac{3\pi}{2}, 2\right)$, and the other extrema in this interval are (0, 3) and (2π, 3).

To find the extrema, we need to find all values of x where the derivative equals zero or where the function is undefined. We start by taking the derivative:

$$y' = \cos(x) + 0$$

$$= \cos(x)$$

Now we set the derivative equal to zero, and solve:

$$0 = \cos(x)$$

$$x = \frac{\pi}{2}, \frac{3\pi}{2}$$

From the unit circle, we see that between 0 and 2π, cos(x) only equals zero when $x = \frac{\pi}{2}$ or $x = \frac{3\pi}{2}$. We also see that the function is defined for all values of x. So the only critical numbers are $x = \frac{\pi}{2}$ and $x = \frac{3\pi}{2}$. We will find the values of the function at each of these critical points and at the two endpoints of the given region (0 and 2π).

At $x = 0$, the function equals:

$$f(0) = \sin(0) + 3$$

$$= 0 + 3$$

$$= 3$$

At $x = \frac{\pi}{2}$, the function equals:

$$f\left(\frac{\pi}{2}\right) = \sin\left(\frac{\pi}{2}\right) + 3$$

$$= 1 + 3$$

$$= 4$$

At $x = \frac{3\pi}{2}$, the function equals:

$$f\left(\frac{3\pi}{2}\right) = \sin\left(\frac{3\pi}{2}\right) + 3$$

$$= -1 + 3$$

$$= 2$$

At $x = 2\pi$, the function equals:

$$f(2\pi) = \sin(2\pi) + 3$$

$$= 0 + 3$$

$$= 3$$

We can see that we have an absolute maximum at $\left(\frac{\pi}{2}, 4\right)$ and an absolute minimum at $\left(\frac{3\pi}{2}, 2\right)$. The other extrema in this interval are (0, 3) and (2π, 3).

6. The correct answer is There is an absolute maximum at $\left(\frac{3\pi}{8}, -0.959\right)$, an absolute minimum at $\left(\frac{\pi}{8}, -1.26\right)$, and the other extrema in this interval are $(0, -1)$ and $\left(\frac{\pi}{2}, -1.22\right)$.

To find the extrema, we need to find all values of x where the derivative equals zero or where the function is undefined. We start by taking the derivative:

$$y' = -\left(-\sin(2x)(2x^{1-1})\right) - (1)\left(\sqrt{2}x^{1-1}\right)$$
$$= 2\sin(2x) - \sqrt{2}$$

Now we set the derivative equal to zero, and solve:

$$0 = 2\sin(2x) - \sqrt{2}$$
$$\sqrt{2} = 2\sin(2x)$$
$$\frac{\sqrt{2}}{2} = \sin(2x)$$
$$2x = \frac{\pi}{4}, \frac{3\pi}{4}$$
$$x = \frac{\pi}{8}, \frac{3\pi}{8}$$

From the unit circle, we see that between 0 and $\frac{\pi}{2}$, $\sin(x)$ only equals zero when $x = \frac{\pi}{8}$ or $x = \frac{3\pi}{8}$. We also see that the function is defined for all values of x. So the only critical numbers are $x = \frac{\pi}{8}$ and $x = \frac{3\pi}{8}$. We will find the values of the function at each of these critical points and at the two endpoints of the given region $\left(0 \text{ and } \frac{\pi}{2}\right)$.

At $x = 0$, the function equals:

$$f(0) = -\cos(2(0)) - \sqrt{2}(0)$$
$$= -1 - 0$$
$$= -1$$

At $x = \frac{\pi}{8}$, the function equals:

$$f\left(\frac{\pi}{8}\right) = -\cos\left(2\left(\frac{\pi}{8}\right)\right) - \sqrt{2}\left(\frac{\pi}{8}\right)$$
$$= -\cos\left(\frac{\pi}{4}\right) - \frac{\pi\sqrt{2}}{8}$$
$$= -\frac{\sqrt{2}}{2} - \frac{\pi\sqrt{2}}{8}$$
$$\approx -1.26$$

Now substitute in $\frac{3\pi}{8}$ for x.

At $x = \frac{3\pi}{8}$, the function equals:

$$f\left(\frac{3\pi}{8}\right) = -\cos\left(2\left(\frac{3\pi}{8}\right)\right) - \sqrt{2}\left(\frac{3\pi}{8}\right)$$
$$= -\cos\left(\frac{3\pi}{4}\right) - \frac{3\pi\sqrt{2}}{8}$$
$$= -\left(-\frac{\sqrt{2}}{2}\right) - \frac{3\pi\sqrt{2}}{8}$$
$$= \frac{\sqrt{2}}{2} - \frac{3\pi\sqrt{2}}{8}$$
$$\approx -0.959$$

At $x = \frac{\pi}{2}$, the function equals:

$$f\left(\frac{\pi}{2}\right) = -\cos\left(2\left(\frac{\pi}{2}\right)\right) - \sqrt{2}\left(\frac{\pi}{2}\right)$$

$$= -\cos(\pi) - \frac{\pi\sqrt{2}}{2}$$

$$= -(-1) - \frac{\pi\sqrt{2}}{2}$$

$$= 1 - \frac{\pi\sqrt{2}}{2}$$

$$\approx -1.22$$

We can see that we have an absolute maximum at $\left(\frac{3\pi}{8}, -0.959\right)$ and an absolute minimum at $\left(\frac{\pi}{8}, -1.26\right)$. The other extrema in this interval are $(0, -1)$ and $\left(\frac{\pi}{2}, -1.22\right)$.

Increasing and decreasing functions

7. The correct answer is the function is increasing for $(-\infty, 0)$, $(0, 0.489)$, and $(3.324, \infty)$, and it is decreasing for $(0.489, 3.324)$.

   We can see that the function will be undefined when $x = 0$. To find the other critical numbers, we will take the derivative and set it equal to zero:

   $$f'(x) = (3)\left(x^{3-1}\right) - (2)\left(5x^{2-1}\right) - (-1)\left(x^{-1-1}\right)$$

   $$= 3x^2 - 10x + \frac{1}{x^2}$$

   $$0 = 3x^2 - 10x + \frac{1}{x^2}$$

   Using a calculator's solver utility, we can find the solutions to this equation to be at $x \approx 0.489$ and $x \approx 3.324$.

So, the critical numbers are $x = 0$, $x \approx 0.489$, and $x \approx 3.324$. These critical numbers separate the function into four different intervals: $(-\infty, 0)$, $(0, 0.489)$, $(0.489, 3.324)$, and $(3.324, \infty)$. To determine if the graph of the function is increasing or decreasing in each interval, we will choose a test point in each region and determine the value of the derivative at that test point.

Region 1, $(-\infty, 0)$:

Test Point: $x = -1$

$$f'(-1) = 3(-1)^2 - 10(-1) + \frac{1}{(-1)^2}$$

$$= 3 + 10 + 1$$

$$= 14$$

Region 2, $(0, 0.489)$:

Test Point: $x = 0.25$

$$f'(0.25) = 3(0.25)^2 - 10(0.25) + \frac{1}{(0.25)^2}$$

$$= 0.1875 - 2.5 + 16$$

$$= 13.6875$$

Region 3, $(0.489, 3.324)$:

Test Point: $x = 1$

$$f'(1) = 3(1)^2 - 10(1) + \frac{1}{(1)^2}$$

$$= 3 - 10 + 1$$

$$= -6$$

Region 4, $(3.324, \infty)$:

Test Point: $x = 10$

$$f'(10) = 3(10)^2 - 10(10) + \frac{1}{(10)^2}$$

$$= 300 - 100 + \frac{1}{100}$$

$$\approx 200$$

Since the derivative (and therefore the slope) is positive in the first, second, and fourth regions, we conclude that the graph is increasing in those regions. Since the derivative is negative in the third region, we conclude that the graph is decreasing in that region.

8. The correct answer is there is a relative maximum at (1.443, 9.808) and a relative minimum at (4.585, −9.808).

   We can see that there are no values of x for which the function is undefined. We need to find the critical numbers, so we will take the derivative and set it equal to zero:

   $$y' = 7\cos(x) + 3\sin(x+2)$$
   $$0 = 7\cos(x) + 3\sin(x+2)$$

   Using a calculator's solver utility, we can find the solutions to this equation to be at $x \approx 1.443$ and $x \approx 4.585$. (Note: There are other solutions, but these are the only two solutions within the requested interval.)

   So the critical numbers are $x \approx 1.443$ and $x \approx 4.585$. These critical numbers separate the function into three different intervals, $(0, 1.443)$, $(1.443, 4.585)$, and $(4.585, 2\pi)$. To determine if the graph of the function is increasing or decreasing in each interval, we will choose a test point in each region and determine the value of the derivative at that test point.

   Region 1, $(0, 1.443)$:

   Test Point: $x = 1$

   $$f'(1) = 7\cos(1) + 3\sin(1+2)$$
   $$\approx 4.205$$

   Region 2, $(1.443, 4.585)$:

   Test Point: $x = 2$

   $$f'(2) = 7\cos(2) + 3\sin(2+2)$$
   $$\approx -5.183$$

   Region 3, $(4.585, 2\pi)$:

   Test Point: $x = 5$

   $$f'(5) = 7\cos(5) + 3\sin(5+2)$$
   $$\approx 3.957$$

   Since the derivative (and therefore the slope) is positive in the first and third regions, we conclude that the graph is increasing in those regions. Since the derivative is negative in the second region, we conclude that the graph is decreasing in that region.

   Because the graph is increasing from 0 to 1.443, then decreasing from 1.443 to 4.585, we can conclude that the graph has a relative maximum at 1.443. Substituting this x value into the function, we get the corresponding y value:

   $$f(1.443) = 7\sin(1.443) - 3\cos(1.443+2)$$
   $$\approx 9.808$$

   So there is a relative maximum at (1.443, 9.808).

   Because the graph is decreasing from 1.443 to 4.585 and increasing from 4.585 to 2π, we can conclude that the graph has a relative minimum at 4.585. Substituting this x value into the function, we get the corresponding y value:

   $$f(4.585) = 7\sin(4.585) - 3\cos(4.585+2)$$
   $$\approx -9.808$$

   So there is a relative minimum at (4.585, −9.808).

9. The correct answer is there is a relative maximum at (2.547, –2.984).

We know from the trigonometric Quotient Identities that the tangent function is equivalent to the sine function divided by the cosine function. So we can rewrite the function as:

$$f(x) = 2\frac{\sin(x)}{\cos(x)} - \frac{1}{2}\frac{\sin\left(\frac{1}{2}x\right)}{\cos\left(\frac{1}{2}x\right)}$$

The function will be undefined whenever $\cos(x) = 0$ and $\cos\left(\frac{1}{2}x\right) = 0$. Using the unit circle, we can see this will occur if $x = \frac{\pi}{2}$, or if $\frac{1}{2}x = \frac{\pi}{2}$. Multiplying the second option through by 2, we can see that the second value occurs when $x = \pi$. So $x = \frac{\pi}{2}$ and $x = \pi$ are critical numbers.

To find the other critical numbers, we will take the derivative and set it equal to zero:

$$f'(x) = 2\sec^2(x) - \frac{1}{2}\left(\frac{1}{2}\right)\sec^2\left(\frac{1}{2}x\right)$$

$$= 2\sec^2(x) - \frac{1}{4}\sec^2\left(\frac{1}{2}x\right)$$

$$0 = 2\sec^2(x) - \frac{1}{4}\sec^2\left(\frac{1}{2}x\right)$$

Using a calculator's solver utility, we can find the solution to this equation to be at $x \approx 2.547$. (Note: There are other solutions, but this is the only solution within the requested interval.)

The critical numbers are therefore $x = \frac{\pi}{2}$, $x \approx 2.547$, and $x = \pi$. These critical numbers and the end points of the

specified region separate the function into three different intervals: $\left(0, \frac{\pi}{2}\right)$, $\left(\frac{\pi}{2}, 2.547\right)$, and $(2.547, \pi)$. To determine if the graph of the function is increasing or decreasing in each interval, we will choose a test point in each region and determine the value of the derivative at that test point.

Region 1, $\left(0, \frac{\pi}{2}\right)$:

Test Point: $x = 1$

$$f'(1) = 2\sec^2(1) - \frac{1}{4}\sec^2\left(\frac{1}{2}(1)\right)$$

$$= 6.526$$

Region 2, $\left(\frac{\pi}{2}, 2.547\right)$:

Test Point: $x = 2$

$$f'(2) = 2\sec^2(2) - \frac{1}{4}\sec^2\left(\frac{1}{2}(2)\right)$$

$$= 10.692$$

Region 3, $(2.547, \pi)$:

Test Point: $x = 3$

$$f'(3) = 2\sec^2(3) - \frac{1}{4}\sec^2\left(\frac{1}{2}(3)\right)$$

$$= -47.92$$

Since the derivative (and therefore the slope) is positive in the first and second regions, we conclude that the graph is increasing in those regions. Since the derivative is negative in the third region, we conclude that the graph is decreasing in that region.

Because the graph is increasing from 0 to 2.547, and then decreasing from 2.547 to π, we can conclude that the graph has a relative maximum at 2.547. Substituting this *x*-value into the function, we get the corresponding *y*-value:

$$f(2.547) = 2\tan(2.547) - \frac{1}{2}\tan\left(\frac{1}{2}(2.547)\right)$$

$$= -2.984$$

There is a relative maximum at (2.547, –2.984).

Chapter Review Solutions

1. The correct answer is $0.5106\dfrac{m^3}{m}$.

First, we need to find an equation that describes the volume of the reservoir as a whole. To obtain this, we will add the expressions for the volume of the sphere and the volume of the cylinder:

$$V_{res} = V_s + V_c$$

$$= \frac{4}{3}\pi r_s^3 + \pi r_c^2 h$$

Now we need to get this expression only in terms of the radius of the sphere. We know the relationship between the radius of the cylinder and the radius of the sphere, so we can substitute this in:

$$r_c = \frac{1}{4}r_s$$

$$V_{res} = \frac{4}{3}\pi r_s^3 + \pi\left(\frac{1}{4}r_s\right)^2 h$$

$$= \frac{4\pi}{3}r_s^3 + \frac{\pi h}{16}r_s^2$$

So now we have an expression for the volume of the reservoir in relation to the radius of the sphere and known constant values. To find $\dfrac{dV_{res}}{dr_s}$, we take the derivative of this expression:

$$\frac{dV_{res}}{dr_s} = (3)\frac{4\pi}{3}r_s^2 + (2)\frac{\pi h}{16}r_s$$

$$= 4\pi r_s^2 + \frac{\pi h}{8}r_s$$

To find the desired rate, we substitute in 0.1 for *h* and 0.2 for r_s:

$$\frac{dV}{dr_s} = 4\pi(0.2)^2 + \frac{\pi(0.1)}{8}(0.2)$$

$$\approx 0.5027 + 0.0079$$

$$\approx 0.5106\frac{m^3}{m}$$

We have given the units as $\dfrac{m^3}{m}$, because what we have found is the rate of change of volume (m^3) with respect to change in radius (m).

2. The correct answer is there is an absolute maximum at $\left(\dfrac{3\pi}{2},\ 4.71\right)$ and an absolute minimum at $\left(\dfrac{5\pi}{6},\ 0.8859\right)$. The other extrema in this interval are (0, 2) and $\left(\dfrac{\pi}{6},\ 2.256\right)$.

To find the extrema, we need to find all values of *x* where the derivative equals zero or where the function is undefined. We start by taking the derivative:

$$f'(x) = 1 - 2\sin(x)$$

Now we set the derivative equal to zero and solve:

$$0 = 1 - 2\sin(x)$$
$$-1 = -2\sin(x)$$
$$\frac{-1}{-2} = \sin(x)$$
$$\sin(x) = \frac{1}{2}$$
$$x = 0, \pi$$

From the unit circle, we see that between 0 and $\frac{3\pi}{2}$, $\sin(x)$ only equals zero when $x = \frac{\pi}{6}$ or when $x = \frac{5\pi}{6}$. We also see that the function is defined for all values of x.

So the only critical numbers are $x = \frac{\pi}{6}$ and $x = \frac{5\pi}{6}$. We will find the values of the function at each of these critical points and at the two endpoints of the given region $\left(0 \text{ and } \frac{3\pi}{2}\right)$.

At $x = 0$, the function equals:

$$f(0) = 0 + 2\cos(0)$$
$$= 2$$

At $x = \frac{\pi}{6}$, the function equals:

$$f\left(\frac{\pi}{6}\right) = \frac{\pi}{6} + 2\frac{\sqrt{3}}{2}$$
$$= \frac{\pi}{6} + \sqrt{3}$$
$$\approx 2.256$$

At $x = \frac{5\pi}{6}$, the function equals:

$$f\left(\frac{5\pi}{6}\right) = \frac{5\pi}{6} + 2\cos\left(\frac{5\pi}{6}\right)$$
$$= \frac{5\pi}{6} - 2\frac{\sqrt{3}}{2}$$
$$= \frac{5\pi}{6} - \sqrt{3}$$
$$= 0.8859$$

Now, substitute in $\frac{3\pi}{2}$ for x.

At $x = \frac{3\pi}{2}$, the function equals:

$$f\left(\frac{3\pi}{2}\right) = \frac{3\pi}{2} + 2\cos\left(\frac{3\pi}{2}\right)$$
$$= \frac{3\pi}{2} + 2(0)$$
$$= \frac{3\pi}{2}$$
$$\approx 4.71$$

We can see that we have an absolute maximum at $\left(\frac{3\pi}{2}, 4.71\right)$ and an absolute minimum at $\left(\frac{5\pi}{6}, 0.8859\right)$. The other extrema in this interval are $(0, 2)$ and $\left(\frac{\pi}{6}, 2.256\right)$.

3. The correct answer is the function is decreasing for (0, 0.552) and increasing for (0.552, ∞).

The function is undefined for $x < 0$ (because then we would have a negative square root). The function is defined for any value of x equal to or greater than zero. We need to find the critical numbers. First, we will rewrite the function in a more convenient form:

$$f(x) = x^4 - x^{\frac{1}{2}}$$

Now, to find the critical numbers, we will take the derivative and set it equal to zero:

$$f'(x) = 4x^3 - \frac{1}{2}x^{-\frac{1}{2}}$$

$$0 = 4x^3 - \frac{1}{2}x^{-\frac{1}{2}}$$

Using a calculator solver, we find that $x \approx 0.552$ is a critical number. This critical number separates the function into two different intervals: (0, 0.552) and (0.552, ∞). To determine if the graph of the function is increasing or decreasing in each interval, we will choose a test point in each region and determine the value of the derivative at that test point.

Region 1, (0, 0.552):

Test Point: $x = 0.5$

$$f'(0.5) = 4(0.5)^3 - \frac{1}{2}(0.5)^{-\frac{1}{2}}$$
$$= 0.5 - 0.707$$
$$= -0.207$$

Region 2, (0.552, ∞):

Test Point: $x = 1$

$$f'(1) = 4(1)^3 - \frac{1}{2}(1)^{-\frac{1}{2}}$$
$$= 4 - 0.5$$
$$= 3.5$$

Since the derivative (and therefore the slope) is negative in the first region, we conclude that the graph is decreasing in that region. Since the derivative is positive in the second region, we conclude that the graph is increasing in that region.

Chapter 8

Integration

In this chapter, we'll review the following concepts:

What is an antiderivative?
Basic integration rules
Particular solutions
Integration by substitution

What is an antiderivative?

An **antiderivative** is the reverse of a derivative. When you are given a derivative function, you can perform the process of **antidifferentiation** to find the original function.

The process of antidifferentiation is also known as **integration.** Put simply, integration undoes differentiation. The two are inverse operations.

The result of integration is an **indefinite integral,** which is always expressed as a function.

The terms antiderivative and indefinite integral mean the same thing.

Many textbooks use the capital letter F to indicate an antiderivative. So, if the derivative function is written as $f(x)$, an antiderivative would be written as $F(x)$.

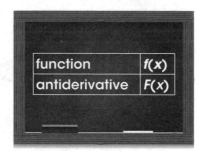

function	$f(x)$
antiderivative	$F(x)$

We can also write the derivative function using prime notation as $f'(x)$. In this case, an antiderivative would be written simply as $f(x)$.

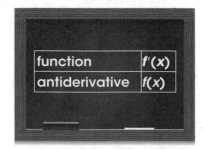

function	$f'(x)$
antiderivative	$f(x)$

The symbol for integration looks like a stretched-out letter s.

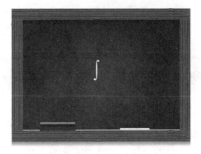

The notation for integration is written as follows:

$$\int f(x)dx = F(x) + c$$

It can also be written using prime notation as shown:

$$\int f'(x)dx = f(x) + c$$

In this case, the function $f(x)$ is the original function. The function $f'(x)$ is the derivative of the original function, and the variable c represents a constant.

All indefinite integrals contain a constant, c.

Here is a diagram showing the parts of the notation:

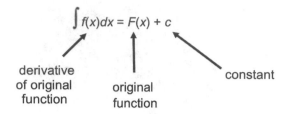

The letters *dx* are used to indicate the variable being integrated.

Here is a diagram containing an actual function:

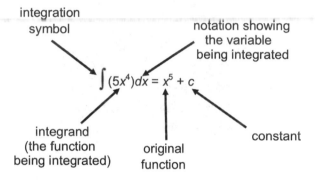

As an example, consider the following function:

$$f(x) = 5x^2 - 2x + 3$$

The derivative of this function is as follows:

$$f'(x) = 10x - 2$$

If we were to integrate the derivative function, we would get the following:

$$\int (10x - 2)dx = 5x^2 - 2x + c$$

In this case, the function produced as the result of integration is the same as the original function, except that we lost the information on the exact value of the constant term. So, we replace this value with the letter *c*. This gives us what is known as a **general solution.**

The only way that we can know the exact value of the constant term is if we are given a point on the original function. We'll see how to find the value of the constant term later in this chapter.

Basic Integration rules

Now that we understand the concept of antiderivatives, let's look at how to find them.

The key rules of integration are given below.

1 The letter *c* must be included in every general solution.

$$\int (n)dx = nx + c$$

The letter *c* is a placeholder for the constant that may be part of the original function.

2 A multiplied constant can be pulled out of the integral.

$$\int nf(x)dx = n\int f(x)dx$$

3 Integrals can be added or subtracted.

$$\int f(x) + g(x)dx = \int f(x)dx + \int g(x)dx$$
$$\int f(x) - g(x)dx = \int f(x)dx - \int g(x)dx$$

To perform these operations, first separate the integrands into two distinct functions. Then integrate them individually, and add or subtract the results.

4 To integrate functions with exponents, use the Power Rule.

$$\int (x^n)dx = \frac{x^{n+1}}{n+1} + c \qquad n \neq -1$$

The Power Rule applies only to exponents other than –1. The function x^{-1} has a special solution that we will not discuss.

Example

Here is an example of integration using the Power Rule.

$$\int (3x^2 - \frac{1}{2}x + 5)dx$$

1 Move through the function term by term.

2 For each term, add 1 to the exponent.

3 Divide the term by the new exponent.

4 Complete steps 2 and 3 for all terms in the function.

$$\int (3x^2 - \frac{1}{2}x + 5)dx = \frac{3x^{2+1}}{2+1} - \frac{1}{2}\left(\frac{x^{1+1}}{1+1}\right) + \frac{5x^{0+1}}{0+1}$$

5 Add *c* to the end.

$$\int (3x^2 - \frac{1}{2}x + 5)dx = \frac{3x^{2+1}}{2+1} - \frac{1}{2}\left(\frac{x^{1+1}}{1+1}\right) + \frac{5x^{0+1}}{0+1} + c$$

Don't forget to add the *c*!

6 Simplify.

$$\int (3x^2 - \frac{1}{2}x + 5)dx = \frac{3x^{2+1}}{2+1} - \frac{1}{2}\left(\frac{x^{1+1}}{1+1}\right) + \frac{5x^{0+1}}{0+1} + C$$

$$= \frac{3x^3}{3} - \frac{1}{2}\left(\frac{x^2}{2}\right) + \frac{5x^1}{1} + C$$

$$= x^3 - \frac{1}{4}x^2 + 5x + C$$

This gives the general solution for the indefinite integral.

Practice Questions—Basic integration rules

Directions: Evaluate the indefinite integrals. You will find the Practice Question Solutions on page 211.

1. $\int (3x^3 - 2x^2 + 1)dx$

2. $\int \frac{1}{x^6}dx$

3. $\int \pi\sqrt{x}dx$

4. $\int e\cos(x)dx$

5. $\int\left(-\frac{\csc(x)\cos(x)}{\sin(x)}\right)dx$

Particular solutions

We can find the value of c for an antiderivative if we are given a point on the original function. This allows us to determine a **particular solution** for an indefinite integral. A particular solution identifies a specific value for c, whereas a general solution does not.

Example 1

In the function we integrated earlier, $f'(x) = 10x - 2$, the general solution was obtained as follows :

$$\int (10x - 2)dx = \frac{10x^{1+1}}{1+1} - \frac{2x^{0+1}}{1} + c$$
$$= \frac{10x^2}{2} - \frac{2x^1}{1} + c$$
$$= 5x^2 - 2x + c$$

If we are given a point on the original function, $f(0) = 3$, we can find the value of c. To do this, we must plug in 0 for x and 3 for $f(x)$ in the general solution:

$$f(x) = 5x^2 - 2x + c$$
$$(3) = 5(0)^2 - 2(0) + c$$
$$3 = 0 - 0 + c$$
$$c = 3$$

The value of c is 3. So, our particular solution for this problem is $5x^2 - 2x + 3$.

Example 2

Here is another example using the function $3x^2 - \frac{1}{2}x + 5$, also integrated earlier. We integrated the function to find the general solution as shown:

$$\int (3x^2 - \frac{1}{2}x + 5)dx = \frac{3x^{2+1}}{2+1} - \frac{1}{2}\left(\frac{x^{1+1}}{1+1}\right) + \frac{5x^{0+1}}{0+1} + c$$
$$= \frac{3x^3}{3} - \frac{1}{2}\left(\frac{x^2}{2}\right) + \frac{5x^1}{1} + c$$
$$= x^3 - \frac{1}{4}x^2 + 5x + c$$

The general solution is $f(x) = x^3 - \frac{1}{4}x^2 + 5x + c$. To find the particular solution, we would need a point on the original function. Say we are given the point $f(0) = 1$. We can plug in 0 for x and 1 for $f(x)$ in the general solution to solve for c:

$$f(x) = x^3 - \frac{1}{4}x^2 + 5x + c$$

$$(1) = (0)^3 - \frac{1}{4}(0)^2 + 5(0) + c$$

$$1 = 0 - 0 + 0 + c$$

$$c = 1$$

The value of c in the general solution is 1. So, our particular solution for this antiderivative is $f(x) = x^3 - \frac{1}{4}x^2 + 5x + 1$.

Differential equations

We can use the knowledge gained about finding particular solutions to solve differential equations. **Differential equations** are equations that involve derivatives. Here is an example.

$$f'' = 18x, \ f'(1) = 16, \ f(0) = 9$$

We have been given the second derivative of a function, as well as two points. One of the points is on the first derivative of the function, $f'(1) = 16$. The second point is on the original function itself, $f(0) = 9$. We are seeking the original function, $f(x)$.

To solve this problem, we take the following steps:

1 Integrate the second derivative to obtain the first derivative:

$$f'(x) = \int f''(x)dx$$

$$= \int 18x \, dx$$

$$= \frac{18x^{1+1}}{1+1} + c$$

$$= \frac{18x^2}{2} + c$$

$$= 9x^2 + c$$

2 Plug in the known value of the first derivative $f'(1) = 16$ to find the value of c:

$$f'(x) = 9x^2 + c$$
$$(16) = 9(1)^2 + c$$
$$(16) = 9(1) + c$$
$$c = 16 - 9$$
$$c = 7$$

This gives us the particular solution for the first derivative:

$$f'(x) = 9x^2 + 7$$

3 Next, integrate the first derivative to obtain the original function:

$$f(x) = \int f'(x)dx$$
$$= \int (9x^2 + 7)dx$$
$$= \frac{9x^{2+1}}{2+1} + \frac{7x^{0+1}}{0+1} + c$$
$$= \frac{9x^3}{3} + \frac{7x^1}{1} + c$$
$$= 3x^3 + 7x + c$$

4 Plug in the known value of the original function, $f(0) = 9$, to find the value of c:

$$f(x) = 3x^3 + 7x + c$$
$$(9) = 3(0)^3 + 7(0) + c$$
$$(9) = 0 + 0 + c$$
$$c = 9$$

This gives us the particular solution for the original function:

$$f(x) = 3x^3 + 7x + 9$$

Practice Questions—Particular solutions

Directions: Follow the directions given to solve the problems below. You will find the Practice Question Solutions on page 212.

Find the general solution of the specified function, and find the particular solution given an initial condition.

6. $f'(x) = \sqrt{x} + 3x + 5$, $f(1) = 2$

7. $g'(x) = \dfrac{2x^3 + 2}{2x^2}$, $g(1) = 7$

Solve the differential equations.

8. $f''(x) = 1$, $f'(2) = 3$, $f(4) = 5$

9. $f''(x) = \sqrt{x}$, $f'(0) = 1$, $f(2) = 3$

Find the requested information.

10. A ball is thrown straight down with an initial velocity of $10\frac{m}{s}$ from the top of a skyscraper 250 meters tall. Find the position function $s(t)$. How long does it take for the ball to strike the ground? Ignore air resistance.

Integration by substitution

Another technique of integration is known as **integration by substitution.** This method can be effective for performing the reverse of differentiation using the Chain Rule. Integration by substitution does not work for every problem. However, it is one of the next steps to try if the basic integration rules do not provide an approach for reaching the solution.

In integration by substitution, we use the variable u to stand in for part of the function to be integrated. For this reason, the approach is sometimes called **u-substitution.**

Here are the steps.

Use u-substitution to find the integral.

$$\int \frac{4x}{(2x^2 + 5)^3} \, dx$$

① Let part of the expression be represented by the variable u.

Let $u = 2x^2 + 5$

The part chosen must be one for which the derivative is also present in the original expression. In this case, the derivative of $2x^2 + 5$ is $4x$, which is present.

2 Take the derivative of u.

$$du = 4xdx$$

3 Replace terms in the original expression using substitution.

$$\int \frac{4x}{(2x^2+5)^3} \, dx = \int \frac{1}{(u)^3} \, du$$

Here we substituted u for $2x^2 + 5$. The term du replaced $4xdx$.

4 Pull any constants out of the integral, if necessary.

There are no constants in this example, so we can skip this step.

5 Perform the integration in terms of u.

$$\int \frac{4x}{(2x^2+5)^3} \, dx = \int \frac{1}{(u)^3} \, du$$
$$= \int u^{-3} \, du$$
$$= \frac{u^{-3+1}}{-3+1} + c$$
$$= \frac{u^{-2}}{-2} + c$$
$$= -\frac{1}{2u^2} + c$$

6 Substitute back in the original expression for u.

$$\frac{1}{2u^2} + c = \frac{1}{2(2x^2+5)^2} + c$$

Here we replaced the u with the original expression, $2x^2 + 5$. The general solution is $\dfrac{1}{2(2x^2+5)^2} + c$.

Example

Let's try another example.

$$\int \frac{3x}{(x^2 + 5)^2} \, dx$$

1 Let part of the expression be represented by the variable *u*.

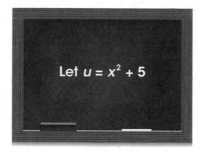

Let $u = x^2 + 5$

The part chosen must be one for which the derivative is also present in the original expression. In this case, the derivative of $x^2 + 5$ is $2x$. The *x*-term is present, but the 2 is not. However, we can accommodate this in step 2.

② Take the derivative of *u*.

$$du = 2xdx$$

In this example, *xdx* is present, so we solve for *xdx*:

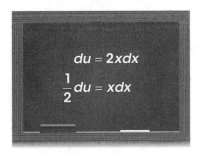

$$du = 2xdx$$

$$\frac{1}{2}du = xdx$$

③ Replace terms in the original expression using substitution.

$$\int \frac{3x}{(x^2 + 5)^2}\, dx = \int \frac{3}{(u)^2}\left(\frac{1}{2}du\right)$$

Here we substituted *u* for $x^2 + 5$. The term $\frac{1}{2}du$ replaced *xdx*.

④ Pull any constants out of the integral if necessary.

$$\int \frac{3x}{(x^2 + 5)^2}\, dx = \int \frac{3}{(u)^2}\left(\frac{1}{2}du\right)$$

$$= \int 3\left(\frac{1}{(u)^2}\right)\left(\frac{1}{2}du\right)$$

$$= \int \frac{3}{2}\left(\frac{1}{(u)^2}\right)du$$

$$= \frac{3}{2}\int\left(\frac{1}{(u)^2}\right)du$$

In this case, we do have one constant, $\frac{3}{2}$. That constant was pulled out and placed before the integral symbol.

⑤ Perform the integration in terms of u.

$$\frac{3}{2}\int\left(\frac{1}{(u)^2}\right)du = \frac{3}{2}\int u^{-2}du$$

$$= \frac{3}{2}\left[\frac{u^{-2+1}}{-2+1}\right] + c$$

$$= \frac{3}{2}\left[\frac{u^{-1}}{-1}\right] + c$$

$$= -\frac{3}{2}\left[\frac{1}{u}\right] + c$$

$$= -\frac{3}{2u} + c$$

⑥ Substitute back in the original expression for u.

$$-\frac{3}{2u} + c = -\frac{3}{2\left(x^2+5\right)} + c$$

$$= -\frac{3}{2x^2+10} + c$$

Here we replaced the u with the original expression, $x^2 + 5$. The general solution is $-\frac{3}{2x^2+10} + c$.

Practice Questions—Integration by substitution

Directions: Use *u*-substitution to find the integral. You will find the Practice Question Solutions on page 215.

11. $\int (x^3 + 2)(3x^2)\,dx$

14. $\int (\cos x \sin x)\,dx$

12. $\int x^4 (x^5 + 5)\,dx$

15. $\int \dfrac{-2x}{(-x^2 + 5)^2}\,dx$

13. $\int x(x+1)^{\frac{1}{3}}\,dx$

Chapter Review

Directions: Follow the directions given to solve the problems below. Solutions can be found on page 218.

Evaluate the indefinite integrals.

1. $\int \left(\frac{1}{2} x^5 + \sqrt{2} x^2 - x - 6 \right) dx$

2. $\int \left(\frac{1}{2x^7} - \frac{1}{3x^3} \right) dx$

3. $\int 36\pi \sin(x) dx$

4. $\int \frac{7 \csc(x)}{\sin(x)} dx$

Find the general solution of the specified function, and find the particular solution given an initial condition.

5. $f'(x) = x^{\frac{3}{2}} + 2x^2 - 4x + 6, \ f(1) = 3$

Solve the differential equations.

6. $f''(x) = x^2, \ f'(4) = 1, \ f(2) = 3$

7. $f''(x) = \frac{1}{x^3}, \ f'(1) = 1, \ f(1) = 1$

Find the requested information.

8. A hockey puck is fired out of a slingshot with an upwards velocity of 15 $\frac{m}{s}$ from the bottom of a 27-meter pit. Find the position function $s(t)$. How long does it take before the puck falls back to the bottom of the pit? Ignore air resistance.

Use *u*-substitution to find the integral.

9. $\int \frac{5}{(2x-1)^3} dx$

10. $\int \sqrt{5x+1} dx$

11. $\int \left(\csc^4 x \cot x \right) dx$

12. $\int \left(4 - x^4 \right) \left(-4x^3 \right) dx$

13. $\int \left(x^2 + 1 \right) \left(x^3 + 3x \right) dx$

14. $\int \left(\sec^2 x \tan x \right) dx$

15. $\int \frac{-6x^2 + 7}{\left(7x - 2x^3 \right)^5} dx$

Practice Question Solutions

Basic integration rules

1. The correct answer is $\frac{3}{4}x^4 - \frac{2}{3}x^3 + x + c$.

   We use the basic integration rules to find the integral:

   $$\int (3x^3 - 2x^2 + 1)\,dx = \frac{3x^{3+1}}{3+1} - \frac{2x^{2+1}}{2+1} + \frac{1x^{0+1}}{0+1} + c$$
   $$= \frac{3}{4}x^4 - \frac{2}{3}x^3 + x + c$$

2. The correct answer is $-\frac{1}{5x^5} + c$.

   First, we rewrite the integral into a more convenient form:

   $$\int \frac{1}{x^6}\,dx = \int x^{-6}\,dx$$

   Now we use the basic integration rules to find the integral of this function:

   $$\int x^{-6}\,dx = \frac{x^{-6+1}}{-6+1} + c$$
   $$= -\frac{x^{-5}}{5} + c$$
   $$= -\frac{1}{5x^5} + c$$

3. The correct answer is $\frac{2\pi}{3}x^{\frac{3}{2}} + c$.

   First, we pull π outside of the integrand since it is a constant:

   $$\int \pi\sqrt{x}\,dx = \pi\int \sqrt{x}\,dx$$

   Next, we rewrite the integral into a more convenient form:

   $$\pi\int \sqrt{x}\,dx = \pi\int x^{\frac{1}{2}}\,dx$$

   Now we use the basic integration rules to find the integral of the function:

   $$\pi\int x^{\frac{1}{2}}\,dx = \pi\left(\frac{x^{\frac{1}{2}+1}}{\frac{1}{2}+1}\right) + c$$
   $$= \pi\left(\frac{x^{\frac{3}{2}}}{\frac{3}{2}}\right) + c$$
   $$= \frac{2\pi}{3}x^{\frac{3}{2}} + c$$

4. The correct answer is $e\sin(x) + c$.

   First, we pull the e outside the integrand since it is a constant:

   $$\int e\cos(x)\,dx = e\int \cos(x)\,dx$$

   Now we take the integral of $\cos(x)$. We can recall that integration is the inverse of differentiation. Since we know the derivative of $\sin(x)$ is $\cos(x)$, going the other direction, we know that the integral of $\cos(x)$ is $\sin(x)$:

   $$e\int \cos(x)\,dx = e\sin(x) + c$$

5.

The correct answer is csc(x) + c.

Particular solutions

6. The correct answer is the general solution is $f(x) = \frac{2}{3}x^{\frac{3}{2}} + \frac{3}{2}x^2 + 5x + c$ and the particular solution is $f(x) = \frac{2}{3}x^{\frac{3}{2}} + \frac{3}{2}x^2 + 5x - \frac{31}{6}$.

To find the general solution, we take the integral of the function:

$$f(x) = \int f'(x)\,dx$$
$$= \int \left(\sqrt{x} + 3x + 5\right)dx$$
$$= \int \left(x^{\frac{1}{2}} + 3x + 5\right)dx$$
$$= \frac{x^{\frac{1}{2}+1}}{\frac{1}{2}+1} + 3\frac{x^{1+1}}{1+1} + 5\frac{x^{0+1}}{0+1} + c$$
$$= \frac{x^{\frac{3}{2}}}{\frac{3}{2}} + 3\frac{x^2}{2} + 5x + c$$
$$= \frac{2}{3}x^{\frac{3}{2}} + \frac{3}{2}x^2 + 5x + c$$

We begin by recognizing from our trigonometric quotient identities that:

$$\frac{\cos(x)}{\sin(x)} = \cot(x)$$

Rewriting the integral, we have:

$$\int(-\csc(x)\cot(x))\,dx$$

Now we take the integral of $-\csc(x)$ $\cot(x)$. We can recall that integration is the inverse of differentiation. Since we know the derivative of csc(x) is $-\csc(x)$ cot(x), going the other direction, we know that the integral of $-\csc(x)\cot(x)$ is csc(x):

$$\int(-\csc(x)\cot(x))\,dx = \csc(x) + c$$

This is our general solution. To find the particular solution, we substitute in the initial condition and solve for c:

$$2 = \frac{2}{3}(1)^{\frac{3}{2}} + \frac{3}{2}(1)^2 + 5(1) + c$$
$$2 = \frac{2}{3} + \frac{3}{2} + 5 + c$$
$$\frac{12}{6} = \frac{4}{6} + \frac{9}{6} + \frac{30}{6} + c$$
$$\frac{12}{6} - \frac{4}{6} - \frac{9}{6} - \frac{30}{6} = c$$
$$-\frac{31}{6} = c$$

And so we can write our particular solution as:

$$f(x) = \frac{2}{3}x^{\frac{3}{2}} + \frac{3}{2}x^2 + 5x - \frac{31}{6}$$

7. The correct answer is the general solution is $g(x) = \frac{1}{2}x^2 - \frac{1}{x} + c$ and the particular solution is $g(x) = \frac{1}{2}x^2 - \frac{1}{x} + \frac{15}{2}$.

To find the general solution, we take the integral of the function:

$$g(x) = \int g'(x)\,dx$$

$$g(x) = \int \left(\frac{2x^3 + 2}{2x^2}\right)dx$$

$$= \int \left(\frac{2x^3}{2x^2} + \frac{2}{2x^2}\right)dx$$

$$= \int \left(x + \frac{1}{x^2}\right)dx$$

$$= \int \left(x + x^{-2}\right)dx$$

$$= \frac{x^{1+1}}{1+1} + \frac{x^{-2+1}}{-2+1} + c$$

$$= \frac{1}{2}x^2 - x^{-1} + c$$

$$= \frac{1}{2}x^2 - \frac{1}{x} + c$$

This is our general solution. To find the particular solution, we substitute in the initial condition and solve for c:

$$7 = \frac{1}{2}(1)^2 - \frac{1}{(1)} + c$$

$$7 = \frac{1}{2} - 1 + c$$

$$7 = \frac{1}{2} - \frac{2}{2} + c$$

$$7 = -\frac{1}{2} + c$$

$$7 + \frac{1}{2} = c$$

$$\frac{14}{2} + \frac{1}{2} = c$$

$$\frac{15}{2} = c$$

And so we can write our particular solution as:

$$g(x) = \frac{1}{2}x^2 - \frac{1}{x} + \frac{15}{2}$$

8. The correct answer is $f(x) = \frac{1}{2}x^2 + x - 7$.

First, we integrate the second derivative to obtain the first derivative:

$$f'(x) = \int f''(x)\,dx$$

$$= \int 1\,dx$$

$$= 1\int dx$$

$$= 1\frac{x^{0+1}}{0+1} + c$$

$$= x + c$$

Now we plug in the known value of the first derivative $f'(2) = 3$ to find the value of c:

$$3 = 2 + c$$

$$1 = c$$

So our first derivative is:

$$f'(x) = x + 1$$

Now we integrate the first derivative to obtain the function:

$$f(x) = \int f'(x)\,dx$$

$$= \int (x + 1)\,dx$$

$$= \frac{x^{1+1}}{1+1} + 1\frac{x^{0+1}}{0+1} + c$$

$$= \frac{1}{2}x^2 + x + c$$

We plug in the known value of the function $f(4) = 5$ to find the value of c:

$$5 = \frac{1}{2}(4)^2 + (4) + c$$

$$5 = 8 + 4 + c$$

$$5 = 12 + c$$

$$-7 = c$$

Finally, we substitute in this value of c to find our particular solution:

$$f(x) = \frac{1}{2}x^2 + x - 7$$

9. The correct answer is

$$f(x) = \frac{4}{15}x^{\frac{5}{2}} + x - 0.5085.$$

First, we integrate the second derivative to obtain the first derivative:

$$f'(x) = \int f''(x)\,dx$$
$$= \int \sqrt{x}\,dx$$
$$= \int x^{\frac{1}{2}}\,dx$$
$$= \frac{x^{\frac{1}{2}+1}}{\frac{1}{2}+1} + c$$
$$= \frac{x^{\frac{3}{2}}}{\frac{3}{2}} + c$$
$$= \frac{2}{3}x^{\frac{3}{2}} + c$$

Now we plug in the known value of the first derivative $f'(0) = 1$ to find the value of c:

$$1 = \frac{2}{3}(0)^{\frac{3}{2}} + c$$
$$1 = c$$

So our first derivative is:

$$f'(x) = \frac{2}{3}x^{\frac{3}{2}} + 1$$

Now we integrate the first derivative to obtain the function:

$$f(x) = \int f'(x)\,dx$$
$$= \int \left(\frac{2}{3}x^{\frac{3}{2}} + 1\right)dx$$
$$= \frac{2}{3}\frac{x^{\frac{3}{2}+1}}{\frac{3}{2}+1} + 1\frac{x^{0+1}}{0+1} + c$$
$$= \frac{2}{3}\frac{x^{\frac{5}{2}}}{\frac{5}{2}} + x + c$$
$$= \frac{4}{15}x^{\frac{5}{2}} + x + c$$

We plug in the known value of the function $f(2) = 3$ to find the value of c:

$$3 = \frac{4}{15}(2)^{\frac{5}{2}} + (2) + c$$
$$\frac{45}{15} = \frac{4}{15}(2)^{\frac{5}{2}} + \frac{30}{15} + c$$
$$\frac{45}{15} - \frac{4}{15}(2)^{\frac{5}{2}} - \frac{30}{15} = c$$
$$1 - \frac{4}{15}(2)^{\frac{5}{2}} = c$$
$$c \approx -0.5085$$

Finally, we substitute in this value of c to find our particular solution:

$$f(x) = \frac{4}{15}x^{\frac{5}{2}} + x - 0.5085$$

10. The correct answer is
$$s(t) = -\frac{9.81}{2}t^2 - 10t + 250 \text{ and } 6.2$$
seconds.

The two initial conditions we were provided are:

$$s(0) = 250$$
$$s'(0) = -10$$

If we assume we are on the Earth, we know the acceleration due to gravity is $-32\,\frac{\text{ft.}}{s^2}$, or $-9.81\,\frac{\text{m}}{s^2}$.

$$a(t) = s''(t)$$
$$= -9.81$$

We can integrate the second derivative of the position function to get the velocity function (which is also the first derivative of the position function):

$$v(t) = s'(t)$$
$$= \int s''(t)\, dt$$
$$= \int -9.81\, dt$$
$$= -9.81 \int dt$$
$$= -9.81 \frac{t^{0+1}}{0+1} + c$$
$$= -9.81t + c$$

To find c, we substitute in the initial condition, $s'(0) = -10$:

$$-10 = -9.81(0) + c$$
$$-10 = c$$

So our velocity function is:

$$s'(t) = -9.81t - 10$$

Now we can integrate the velocity function to obtain the position function:

$$s(t) = \int s'(t)\, dt$$
$$= \int (-9.81t - 10)\, dt$$
$$= -9.81 \frac{t^{1+1}}{1+1} - 10 \frac{t^{0+1}}{0+1} + c$$
$$= -\frac{9.81}{2}t^2 - 10t + c$$

To find c, we substitute in the initial condition, $s(0) = 250$:

$$250 = -\frac{9.81}{2}(0)^2 - 10(0) + c$$
$$250 = c$$

So our position function is:

$$s(t) = -\frac{9.81}{2}t^2 - 10t + 250$$

To find the time when the ball strikes the ground, we substitute 0 for position into the position function and solve for time:

$$0 = -\frac{9.81}{2}t^2 - 10t + 250$$

$$t = \frac{-(-10) \pm \sqrt{(-10)^2 - 4\left(-\frac{9.81}{2}\right)(250)}}{2\left(-\frac{9.81}{2}\right)}$$

$$= \frac{10 \pm \sqrt{100 + 4905}}{-9.81}$$

$$= -8.2,\ 6.2$$

Selecting the positive answer, we conclude that the ball strikes the ground 6.2 seconds after it is thrown.

Integration by substitution

11. The correct answer is $\frac{x^6}{2} + 2x^3 + c$.

   We recognize that if we let the $x^3 + 2$ term equal u, its derivative is $3x^2$, which is present.

   $$u = x^3 + 2$$
   $$du = 3x^2 dx$$

   We use this approach to do the u-substitution:

   $$\int (x^3 + 2)(3x^2)\, dx = \int (u)\, du$$
   $$= \frac{u^2}{2} + c$$

Now we replace u to arrive at our solution:

$$\frac{u^2}{2} + c = \frac{\left(x^3 + 2\right)^2}{2} + c$$

$$= \frac{x^6 + 4x^3 + 4}{2} + c$$

$$= \frac{x^6}{2} + 2x^3 + 2 + c$$

$$= \frac{x^6}{2} + 2x^3 + c$$

You might notice that we removed the +2 term. Since we are already adding a constant c, there is no need to add 2 to this function, and it can be removed.

12. The correct answer is $\frac{1}{10}x^{10} + x^5 + \frac{5}{2} + c$.

We recognize that if we let the $x^5 + 5$ term equal u, its derivative is $5x^4$. The x^4 term is present, but the 5 is missing. When we solve for du, we see that we can accommodate this when we do the u-substitution:

$$u = x^5 + 5$$

$$du = 5x^4 dx$$

$$\frac{1}{5} du = x^4 dx$$

We can now do the u-substitution and solve the integral:

$$\int x^4 \left(x^5 + 5\right) dx = \frac{1}{5}\int u\, du$$

$$= \frac{u^2}{(5)(2)} + c$$

$$= \frac{u^2}{10} + c$$

Now we replace u to arrive at our solution:

$$\frac{u^2}{10} + c = \frac{\left(x^5 + 5\right)^2}{10} + c$$

$$= \frac{x^{10} + 10x^5 + 25}{10} + c$$

$$= \frac{1}{10}x^{10} + x^5 + \frac{5}{2} + c$$

13. The correct answer is

$$\frac{3}{7}(x+1)^{\frac{7}{3}} - \frac{3}{4}(x+1)^{\frac{4}{3}} + c\,.$$

We will let $u = x + 1$ and change the variables in the function from x to u:

$$u = x + 1$$

$$du = dx$$

$$u - 1 = x$$

We use this approach to do the u-substitution:

$$\int x\left(x+1\right)^{\frac{1}{3}} dx = \int (u-1)u^{\frac{1}{3}} du$$

$$= \int uu^{\frac{1}{3}} - u^{\frac{1}{3}} du$$

$$= \int u^{\frac{4}{3}} - u^{\frac{1}{3}} du$$

$$= \frac{u^{\frac{7}{3}}}{\frac{7}{3}} - \frac{u^{\frac{4}{3}}}{\frac{4}{3}} + c$$

$$= \frac{3}{7}u^{\frac{7}{3}} - \frac{3}{4}u^{\frac{4}{3}} + c$$

Now we replace u to arrive at our solution.

$$\frac{3}{7}u^{\frac{7}{3}} - \frac{3}{4}u^{\frac{4}{3}} + c = \frac{3}{7}(x+1)^{\frac{7}{3}} - \frac{3}{4}(x+1)^{\frac{4}{3}} + c$$

14. The correct answer is $\frac{1}{2}\sin^2(x) + c$.

    We will let $u = \sin(x)$. We can see that the derivative of $\sin(x)$, which is $\cos(x)$, is also present.

    $$u = \sin(x)$$
    $$du = \cos(x)dx$$

    We use this approach to do the u-substitution and solve the integral:

    $$\int(\cos x \sin x)dx = \int u\,du$$
    $$= \frac{u^2}{2} + c$$

    Now we replace u to arrive at our solution:

    $$\frac{u^2}{2} + c = \frac{1}{2}\sin^2(x) + c$$

15. The correct answer is $-\frac{1}{5 - x^2} + c$.

    We will let $u = -x^2 + 5$. We can see that the derivative of this is present.

    $$u = -x^2 + 5$$
    $$du = -2x\,dx$$

    We use this approach to do the u-substitution and solve the integral:

    $$\int \frac{-2x}{\left(-x^2 + 5\right)^2}dx = \int \frac{1}{u^2}du$$
    $$= \int u^{-2}du$$
    $$= \frac{u^{-1}}{-1} + c$$
    $$= -\frac{1}{u} + c$$

    Now we replace u to arrive at our solution:

    $$-\frac{1}{u} + c = -\frac{1}{-x^2 + 5} + c$$
    $$= -\frac{1}{5 - x^2} + c$$

Chapter Review Solutions

1. The correct answer is $\frac{1}{12}x^6 + \frac{\sqrt{2}}{3}x^3 - \frac{1}{2}x^2 - 6x + c$.

We use the basic integration rules to find the integral:

$$\int\left(\frac{1}{2}x^5 + \sqrt{2}x^2 - x - 6\right)dx = \left(\frac{1}{2}\right)\frac{x^{5+1}}{(5+1)} + \sqrt{2}\frac{x^{2+1}}{(2+1)} - \frac{x^{1+1}}{(1+1)} - 6\frac{x^{0+1}}{(0+1)} + c$$

$$= \frac{1}{12}x^6 + \frac{\sqrt{2}}{3}x^3 - \frac{1}{2}x^2 - 6x + c$$

2. The correct answer is $-\frac{1}{12x^6} + \frac{1}{6x^2} + c$.

First, we rewrite the integral into a more convenient form:

$$\int\left(\frac{1}{2x^7} - \frac{1}{3x^3}\right)dx = \int\frac{1}{2x^7}dx - \int\frac{1}{3x^3}dx$$

$$= \frac{1}{2}\int x^{-7}dx - \frac{1}{3}\int x^{-3}dx$$

Now we use the basic integration rules to find the integral:

$$\frac{1}{2}\int x^{-7}dx - \frac{1}{3}\int x^{-3}dx = \left(\frac{1}{2}\right)\frac{x^{-7+1}}{(-7+1)} - \left(\frac{1}{3}\right)\frac{x^{-3+1}}{-3+1} + c$$

$$= -\frac{1}{12}x^{-6} + \frac{1}{6}x^{-2} + c$$

$$= -\frac{1}{12x^6} + \frac{1}{6x^2} + c$$

3. The correct answer is $-36\pi\cos(x) + c$.

First, we pull the 36π outside the integrand, since it is a constant:

$$\int 36\pi\sin(x)dx = 36\pi\int\sin(x)dx$$

Now we take the integral of $\sin(x)$. We can recall that integration is the inverse of differentiation. Since we know the derivative of $\cos(x)$ is negative $\sin(x)$, going the other direction, we know that the integral of negative $\sin(x)$ is $\cos(x)$. We are missing a negative sign, so we multiply by (-1) twice to obtain it:

$$36\pi\int\sin(x)dx = -36\pi\int\left(-\sin(x)\right)dx$$

Now we can take the integral.

$$-36\pi \int \left(-\sin(x)\right) dx = -36\pi \cos(x) + c$$

4. The correct answer is $-7\cot(x) + c$.

We begin by recognizing from our trig identities that the reciprocal of the sine function is the cosecant function:

$$\frac{1}{\sin(x)} = \csc(x)$$

Rewriting the integral, we have:

$$\int \frac{7\csc(x)}{\sin(x)} dx = \int 7\csc(x)\csc(x) dx$$
$$= 7\int \csc^2(x) dx$$

Now we take the integral of $\csc^2(x)$. We can recall that integration is the inverse of differentiation. Since we know the derivative of the $\cot(x) = -\csc^2(x)$, going the other direction, we know that the integral of $-\csc^2(x)$ is $\cot(x)$. We are missing a negative sign, so we multiply by (-1) twice to obtain it, and then solve the integral:

$$7\int \csc^2(x) dx = -7\int \left(-\csc^2(x)\right) dx$$
$$= -7\cot(x) + c$$

5. The correct answer is The general solution is $f(x) = \frac{2}{5}x^{\frac{5}{2}} + \frac{2}{3}x^3 - 2x^2 + 6x + c$ and the particular solution is $f(x) = \frac{2}{5}x^{\frac{5}{2}} + \frac{2}{3}x^3 - 2x^2 + 6x - \frac{31}{15}$.

To find the general solution, we take the integral of the function:

$$f(x) = \int f'(x) dx$$
$$= \int \left(x^{\frac{3}{2}} + 2x^2 - 4x + 6\right) dx$$
$$= \frac{x^{\frac{3}{2}+1}}{\left(\frac{3}{2} + 1\right)} + 2\frac{x^{2+1}}{(2+1)} - 4\frac{x^{1+1}}{(1+1)} + 6\frac{x^{0+1}}{(0+1)} + c$$
$$= \frac{2}{5}x^{\frac{5}{2}} + \frac{2}{3}x^3 - 2x^2 + 6x + c$$

This is our general solution. To find the particular solution, we substitute in the initial condition and solve for c:

$$3 = \frac{2}{5}(1)^{\frac{5}{2}} + \frac{2}{3}(1)^3 - 2(1)^2 + 6(1) + c$$
$$3 = \frac{2}{5} + \frac{2}{3} + 4 + c$$
$$\frac{45}{15} = \frac{6}{15} + \frac{10}{15} + \frac{60}{15} + c$$
$$c = \frac{45}{15} - \frac{6}{15} - \frac{10}{15} - \frac{60}{15}$$
$$= -\frac{31}{15}$$

And so we can write our particular solution as:

$$f(x) = \frac{2}{5}x^{\frac{5}{2}} + \frac{2}{3}x^3 - 2x^2 + 6x - \frac{31}{15}$$

6. The correct answer is $f(x) = \frac{1}{12}x^4 - \frac{61}{3}x + \frac{127}{3}$.

First, we integrate the second derivative to obtain the first derivative:

$$f'(x) = \int f''(x)\,dx$$
$$= \int x^2\,dx$$
$$= \frac{x^3}{3} + c$$

Now we plug in the known value of the first derivative $f'(4) = 1$ to find the value of c:

$$1 = \frac{4^3}{3} + c$$
$$1 = \frac{64}{3} + c$$
$$c = 1 - \frac{64}{3}$$
$$c = \frac{3}{3} - \frac{64}{3}$$
$$c = -\frac{61}{3}$$

So our first derivative is:

$$f'(x) = \frac{x^3}{3} - \frac{61}{3}$$

Now we integrate the first derivative to obtain the function:

$$f(x) = \int f'(x)\,dx$$
$$= \int \left(\frac{x^3}{3} - \frac{61}{3} \right) dx$$
$$= \frac{x^{3+1}}{3(3+1)} - \frac{61}{3}\frac{x^{0+1}}{(0+1)} + c$$
$$= \frac{1}{12}x^4 - \frac{61}{3}x + c$$

We plug in the known value of the function $f(2) = 3$ to find the value of c:

$$3 = \frac{1}{12}(2)^4 - \frac{61}{3}(2) + c$$
$$3 = \frac{16}{12} - \frac{122}{3} + c$$
$$c = 3 - \frac{16}{12} + \frac{122}{3}$$
$$= \frac{9}{3} - \frac{4}{3} + \frac{122}{3}$$
$$= \frac{127}{3}$$

Finally, we substitute in this value of c to find our particular solution:

$$f(x) = \frac{1}{12}x^4 - \frac{61}{3}x + \frac{127}{3}$$

7. The correct answer is $f(x) = \frac{1}{2}x^{-1} + \frac{3}{2}x - 1$.

First, we integrate the second derivative to obtain the first derivative:

$$f'(x) = \int f''(x)\,dx$$
$$= \int \frac{1}{x^3}\,dx$$
$$= \int x^{-3}\,dx$$
$$= \frac{x^{-3+1}}{-3+1} + c$$
$$= -\frac{1}{2}x^{-2} + c$$

Now we plug in the known value of the first derivative $f'(1) = 1$ to find the value of c:

$$1 = -\frac{1}{2}(1)^{-2} + c$$
$$1 = -\frac{1}{2} + c$$
$$1 + \frac{1}{2} = c$$
$$c = \frac{3}{2}$$

So our first derivative is as shown:

$$f'(x) = -\frac{1}{2}x^{-2} + \frac{3}{2}$$

Now we integrate the first derivative to obtain the function:

$$f(x) = \int f'(x)dx$$
$$= \int \left(-\frac{1}{2}x^{-2} + \frac{3}{2}\right)dx$$
$$= \int -\frac{1}{2}x^{-2}dx + \int \frac{3}{2}dx$$
$$= -\frac{1}{2}\int x^{-2}dx + \frac{3}{2}\int dx$$
$$= -\frac{1}{2}\frac{x^{-2+1}}{(-2+1)} + \frac{3}{2}\frac{x^{0+1}}{(0+1)} + c$$
$$= \frac{1}{2}x^{-1} + \frac{3}{2}x + c$$

Now we plug in the known value of the function $f(1) = 1$ to find the value of c:

$$1 = \frac{1}{2}(1)^{-1} + \frac{3}{2}(1) + c$$
$$1 = \frac{1}{2} + \frac{3}{2} + c$$
$$\frac{2}{2} - \frac{1}{2} - \frac{3}{2} = c$$
$$c = -1$$

Finally, we substitute in this value of c to find our particular solution:

$$f(x) = \frac{1}{2}x^{-1} + \frac{3}{2}x - 1$$

8. The correct answer is $s(t) = -\frac{9.81}{2}t^2 + 15t - 27$ and 3.058 seconds.

The two initial conditions we were provided are the following:

$$s(0) = -27$$
$$s'(0) = 15$$

Since we are on Earth, we know the acceleration due to gravity is $-9.81\frac{m}{s^2}$.

$$a(t) = s''(t)$$
$$= -9.81$$

We can integrate the second derivative of the position function to get the velocity function:

$$v(t) = s'(t)$$
$$= \int s''(t)$$
$$= \int -9.81dt$$
$$= -9.81\int dt$$
$$= -9.81\frac{t^{0+1}}{0+1} + c$$
$$= -9.81t + c$$

To find c, we substitute in the initial condition, $s'(0) = 15$:

$$15 = -9.81(0) + c$$
$$15 = c$$

So, our velocity function is:

$$s'(t) = -9.81t + 15$$

Now we can integrate the velocity function to obtain the position function:

$$s(t) = \int s'(t)\,dt$$

$$= \int (-9.81t + 15)\,dt$$

$$= -9.81\frac{t^{1+1}}{1+1} + 15\frac{t^{0+1}}{0+1} + c$$

$$= -\frac{9.81}{2}t^2 + 15t + c$$

To find c, we substitute in the initial condition, $s(0) = -27$:

$$-27 = -\frac{9.81}{2}(0)^2 + 15(0) + c$$

$$-27 = c$$

So our position function is:

$$s(t) = -\frac{9.81}{2}t^2 + 15t - 27$$

To find out how long it takes for the puck to fall back to the bottom of the pit, we substitute –27 for position into the position function and solve for time:

$$-27 = -\frac{9.81}{2}t^2 + 15t - 27$$

$$0 = -\frac{9.81}{2}t^2 + 15t$$

$$0 = t\left(-\frac{9.81}{2}t + 15\right)$$

$$0 = -\frac{9.81}{2}t + 15$$

$$-15 = -\frac{9.81}{2}t$$

$$t = (-15)\left(-\frac{2}{9.81}\right)$$

$$\approx 3.058$$

and

$$t = 0$$

We recognize that when $t = 0$, we are at our initial condition when the puck is first

being launched with the slingshot. The second value, $t = 3.058$ seconds, is when the puck returns to the bottom of the pit.

9. The correct answer is $-\dfrac{5}{4(2x-1)^2} + c$.

We recognize that if we let the $2x - 1$ term equal u, its derivative is 2. However, when we solve for du, we see that we can accommodate this when we do the u-substitution:

$$u = 2x + 1$$

$$du = 2dx$$

$$\frac{1}{2}du = dx$$

We can now do the u-substitution and solve the integral:

$$\int \frac{5}{(2x-1)^3}\,dx = \int \left[\frac{5}{(u)^3}\right]\frac{1}{2}\,du$$

$$= \int \frac{5}{2u^3}\,du$$

$$= \frac{5}{2}\int \frac{1}{u^3}\,du$$

$$= \frac{5}{2}\int u^{-3}\,du$$

$$= \frac{5}{2}\left[\frac{u^{-3+1}}{-3+1}\right] + c$$

$$= \frac{5}{2}\left[\frac{u^{-2}}{-2}\right] + c$$

$$= -\frac{5}{4}u^{-2} + c$$

$$= -\frac{5}{4u^2} + c$$

Now we replace u to arrive at our solution:

$$-\frac{5}{4u^2} + c = -\frac{5}{4(2x-1)^2} + c$$

10. The correct answer is $\dfrac{2(5x+1)^{\frac{3}{2}}}{15} + c$.

We recognize that if we let the $5x + 1$ term equal u, its derivative is 5. However, when we solve for du, we see that we can accommodate this when we do the u-substitution:

$$u = 5x + 1$$
$$du = 5dx$$
$$\frac{1}{5}du = dx$$

We can now do the u-substitution and solve the integral:

$$\int \sqrt{5x+1}\, dx = \int \sqrt{u}\, \frac{1}{5}du$$
$$= \frac{1}{5}\int \sqrt{u}\, du$$
$$= \frac{1}{5}\int \left(u^{\frac{1}{2}}\right) du$$
$$= \frac{1}{5}\left[\frac{u^{\frac{1}{2}+1}}{\frac{1}{2}+1}\right] + c$$
$$= \frac{1}{5}\left[\frac{u^{\frac{3}{2}}}{\frac{3}{2}}\right] + c$$
$$= \frac{1}{5}\left(\frac{2}{3}\right)\left(u^{\frac{3}{2}}\right) + c$$
$$= \frac{2}{15}\left(u^{\frac{3}{2}}\right) + c$$
$$= \frac{2u^{\frac{3}{2}}}{15} + c$$

Now we replace u to arrive at our solution:

$$\frac{2u^{\frac{3}{2}}}{15} + c = \frac{2(5x+1)^{\frac{3}{2}}}{15} + c$$

11. The correct answer is $-\dfrac{1}{4}\csc^4(x) + c$.

If we let $u = \csc(x)$, the derivative $\csc(x)\cot(x)$ is present.

$$u = \csc(x)$$
$$du = -\csc(x)\cot(x)\,dx$$
$$-du = \csc(x)\cot(x)\,dx$$

We use u-substitution to find the integral:

$$\int \csc^4(x)\cot(x)\,dx = \int \csc^3(x)\csc(x)\cot(x)\,dx$$
$$= \int u^3(-du)$$
$$= -1\int u^3\,du$$
$$= -\left[\frac{u^{3+1}}{3+1}\right] + c$$
$$= -\left[\frac{u^4}{4}\right] + c$$
$$= -\left[\frac{1}{4}u^4\right] + c$$
$$= -\frac{1}{4}u^4 + c$$

Now, replace u to arrive at the solution:

$$-\frac{1}{4}u^4 + c = -\frac{1}{4}[\csc(x)]^4 + c$$
$$= -\frac{1}{4}\csc^4(x) + c$$

12. The correct answer is $\dfrac{1}{2}x^8 - 4x^4 + c$.

We recognize that if we let the $4 - x^4$ term equal u, its derivative is $-4x^3$, which is present.

$$u = 4 - x^4$$
$$du = -4x^3 dx$$

We use this approach to do the u-substitution:

$$\int \left(4 - x^4\right)\left(-4x^3\right)dx = \int u\,du$$
$$= \frac{u^2}{2} + c$$

Now we replace u to arrive at our solution:

$$\frac{u^2}{2} + c = \frac{\left(4 - x^4\right)^2}{2} + c$$
$$= \frac{16 - 8x^4 + x^8}{2} + c$$
$$= \frac{16}{2} - \frac{8}{2}x^4 + \frac{1}{2}x^8 + c$$
$$= \frac{1}{2}x^8 - 4x^4 + 8 + c$$
$$= \frac{1}{2}x^8 - 4x^4 + c$$

You might notice that we removed the +8 term. Since we are already adding a constant c, there is no need to add 8 to this function, and it can be removed.

13. The correct answer is $\frac{1}{6}x^6 + x^4 + \frac{3}{2}x^2 + c$.

We recognize that if we let the $x^3 + 3x$ term equal u, its derivative is $3x^2 + 3$. Something very close to this is present. If we were to multiply the $x^2 + 1$ term by 3, we would have the needed derivative. When we solve for du, we see that we can accommodate this when we do the u-substitution:

$$u = x^3 + 3x$$
$$du = \left(3x^2 + 3\right)dx$$
$$du = 3\left(x^2 + 1\right)dx$$
$$\frac{1}{3}du = \left(x^2 + 1\right)dx$$

We can now do the u-substitution and solve the integral:

$$\int \left(x^2 + 1\right)\left(x^3 + 3x\right)dx = \frac{1}{3}\int u\,du$$
$$= \frac{1}{3}\frac{u^2}{2} + c$$
$$= \frac{1}{6}u^2 + c$$

Now we replace u to arrive at our solution:

$$\frac{1}{6}u^2 + c = \frac{1}{6}\left(x^3 + 3x\right)^2 + c$$
$$= \frac{1}{6}\left(x^6 + 6x^4 + 9x^2\right) + c$$
$$= \frac{1}{6}x^6 + x^4 + \frac{3}{2}x^2 + c$$

14. The correct answer is $\frac{1}{2}\sec^2(x) + c$.

We can rewrite the integral as:

$$\int \left(\sec^2(x)\tan(x)\right)dx =$$
$$\int \left(\sec(x)\sec(x)\tan(x)\right)dx$$

Now we can see that if we let $u = \sec(x)$, the derivative of $\sec(x)$ is $\sec(x)\tan(x)$, which is present.

$$u = \sec(x)$$
$$du = \sec(x)\tan(x)dx$$

We use this approach to do the u-substitution and solve the integral:

$$\int \left(\sec(x)\sec(x)\tan(x)\right)dx = \int u\,du$$
$$= \frac{u^2}{2} + c$$

Now we replace u to arrive at our solution:

$$\frac{u^2}{2} + c = \frac{1}{2}\sec^2(x) + c$$

15. The correct answer is $-\dfrac{1}{4\left(7x-2x^3\right)^4}+c$.

We will let $u = 7x - 2x^3$. We can see the derivative of this is present:

$$u = 7x - 2x^3$$
$$du = \left(7 - 6x^2\right)dx$$
$$du = \left(-6x^2 + 7\right)dx$$

We use this approach to do the u-substitution and solve the integral:

$$\int \frac{-6x^2 + 7}{\left(7x - 2x^3\right)^5}dx = \int \frac{1}{u^5}du$$
$$= \int u^{-5}du$$
$$= \frac{u^{-4}}{-4} + c$$
$$= -\frac{1}{4u^4} + c$$

Now we replace u to arrive at our solution:

$$-\frac{1}{4u^4} + c = -\frac{1}{4\left(7x - 2x^3\right)^4} + c$$

Chapter 9

Definite Integrals

In this chapter, we'll review the following concepts:

What is a definite integral?
Finding area
The Fundamental Theorem of Calculus
Finding definite integrals

What is a definite integral?

A **definite integral** differs from an indefinite integral in two main respects. First, a definite integral has specific endpoints, whereas an indefinite integral does not. Second, the result of finding a definite integral is always a number, while indefinite integrals always result in functions.

Here is the form for a definite integral.

$$\int_a^b f(x)dx$$

The variables a and b represent numbers that mark the beginning and end of the interval being integrated. These numbers are x-values on the graph of the function. For instance, let's say we have a definite integral like this:

$$\int_2^4 (5x^2)dx$$

The notation tells us to integrate the function $f(x) = 5x^2$ on the interval between $x = 2$ and $x = 4$.

The top and bottom x-values are called the **limits of integration**.

egghead's Guide to Calculus

Definite integrals can be solved using many of the same rules we used for indefinite integrals. One additional rule may also be useful to know:

$$\int_a^b f(x)dx = -\int_b^a f(x)dx$$

This rule states that we can switch the order of the top and bottom limits and add a minus (negative) sign before the integral.

Finding area

Definite integrals help us solve a specific type of problem, known as the **area problem.** Simply put, how do we find the area under a curve? As an example, consider the curve shown:

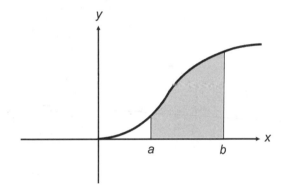

If we wanted to find the area of the shaded portion between the *x*-values of *a* and *b*, one way to do so would be to divide this area into rectangles. We could compute the area of each rectangle and add the areas together.

This approach would only give us an estimation of the area under the curve. If we were finding the area under a straight line, the area determined using rectangles would be exact:

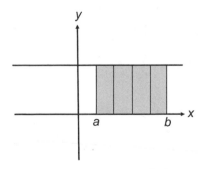

But because many functions are curves, we get only an approximation when we use rectangles.

Whether the result is an overestimation or an underestimation depends on the shape of the curve. It also depends on whether we use the right or left sides of the rectangles to measure their height.

In the figure below, we have divided the shaded portion into 4 rectangles. The right sides of the rectangles are used to indicate the rectangles' heights. The rectangles extend slightly above the top of the curve. If we added their areas together, the total would overestimate the area under the curve by a small amount.

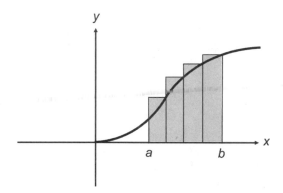

If we drew the rectangles another way, using their left sides to indicate height, we'd still only have an estimate of the area between *a* and *b*. This time, the area would be an underestimate:

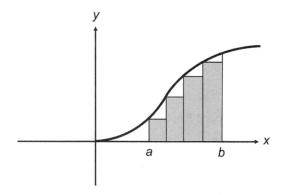

Even though these areas aren't exact, finding the sums of the areas of rectangles is one way to get an answer that is close to the actual area under the curve. We will explore two techniques that allow us to determine the area in this fashion.

Using sums

Since we will be adding together multiple values, we call this process summation. In this case, **summation** involves taking the sum of many areas to determine the area estimate.

In the big picture, the process of adding up smaller areas allows us to approximate the area under the curve:

$$
\begin{array}{l}
\text{approximate} \\
\text{area under} \\
\text{curve}
\end{array}
=
\begin{array}{l}
\text{area of} \\
\text{rectangle 1}
\end{array}
+
\begin{array}{l}
\text{area of} \\
\text{rectangle 2}
\end{array}
+
\begin{array}{l}
\text{area of} \\
\text{rectangle 3}
\end{array}
+
\begin{array}{l}
\text{area of} \\
\text{rectangle 4}
\end{array}
\; \ldots \text{and so forth}
$$

We use a certain notation when working with sums.

This notation, called **summation notation,** starts with the Greek letter sigma:

An example of an operation given in summation notation might be as follows:

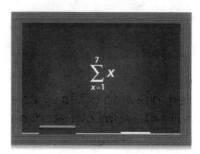

In this formula, the variable x is used to identify the numbers given. The value $x = 1$ tells us the number to use as a starting point. The number 7 is the number to use as the ending point. The variable x, to the right of the sigma, indicates the operation to be performed on each number.

The sigma symbol indicates to add up the results.

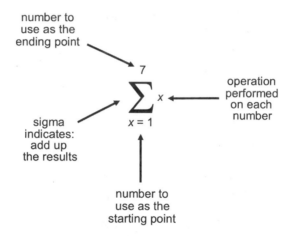

The notation above tells us to add up the numbers from 1 to 7:

$$\sum_{x=1}^{7} x = 1 + 2 + 3 + 4 + 5 + 6 + 7$$
$$= 28$$

Here is another example:

$$\sum_{x=3}^{6} x^2$$

This notation instructs us to add up the squares of the numbers between 3 and 6:

$$\sum_{x=3}^{6} x^2 = (3)^2 + (4)^2 + (5)^2 + (6)^2$$
$$= 9 + 16 + 25 + 36$$
$$= 86$$

In summation notation, variables other than x may be used. The letter i is often used, for example.

$$\sum_{i=3}^{6} i^2$$

The letter i is referred to as the **index**.

The following formulas may be helpful when computing sums. In these formulas, the letter n is used to represent the ending point.

Summation with a constant, c	$\sum_{i=1}^{n} c = cn$
Values of i not raised to a power	$\sum_{i=1}^{n} i = \dfrac{n(n+1)}{2}$
Values of i raised to the second power	$\sum_{i=1}^{n} i^2 = \dfrac{n(n+1)(2n+1)}{6}$
Values of i raised to the third power	$\sum_{i=1}^{n} i^3 = \dfrac{n^2(n+1)^2}{4}$

Area estimation

The technique of summation is important for adding together the areas of large numbers of rectangles to estimate the area under a curve.

To estimate the area under a curve using rectangles, we take the following steps:

1 Divide the area into rectangles, each with a width of 1.

2 Measure the height on the left or right side of each rectangle.

3 Calculate the area of each rectangle.

4 Add up the areas.

Here is an example.

Use four rectangles to find an approximation for the area between the graph of $y = x^3$ and the x axis between $x = 0$ and $x = 4$.

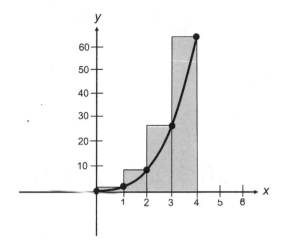

Note: Figure not drawn to scale.

To understand this problem, we graph the function between $x = 0$ and $x = 4$, as shown in the figure. We then divide the region into four different sections, each with a width of 1, and create a rectangle in each section. We can find the height of each rectangle by realizing that the rectangles we have drawn have their highest points at their right edges. We can solve the function at each of these edges to find the height of each of these rectangles.

$$f(1) = (1)^3$$
$$= 1$$

$$f(2) = (2)^3$$
$$= 8$$

$$f(3) = (3)^3$$
$$= 27$$

$$f(4) = (4)^3$$
$$= 64$$

Now we can find the approximation for the area by finding the area of each rectangle and adding them together:

$$\text{Area} = (1)(1) + (8)(1) + (27)(1) + (64)(1)$$
$$= 100$$

We can see from graphing the function and the rectangles on the same graph that in this case, this approximation overestimates the area. Other approximations are possible.

Practice Questions—Finding area

Directions: Follow the directions given to solve the problems below. You will find the Practice Question Solutions on page 245.

Find the indicated sum.

1. $\sum_{k=1}^{3}\left(k^2 - k + 1\right)$

2. $\sum_{j=1}^{n}\dfrac{j^2 + j}{n+1}$, for $n = 2, 5, 12$

3. $\sum_{i=1}^{n}\left(\dfrac{1}{2}i^3 + 5\right)$, for $n = 1, 10$

Find the indicated area.

4. Use three rectangles to find an approximation for the area between the graph of $y = x^3 + 3$ and the x-axis between $x = 0$ and $x = 3$.

5. Use four rectangles to find an approximation for the area between the graph of $y = \sqrt{x} + 1$, $y = 1$, $x = 0$ and $x = 4$.

The Fundamental Theorem of Calculus

In the previous section, we used the sums of the areas of rectangles to estimate the area under a curve. The results obtained were only approximations, as we saw. We could use other methods of estimating the area and even use other figures, such as trapezoids, which might give us an answer closer to the actual area under the curve. To get the exact area using rectangles, however, we would have to increase the number of rectangles used. If we did so, the width of each rectangle would decrease.

We would wind up with many very narrow rectangles

As the number of rectangles approached infinity, and the width of the rectangles approached zero, we would get a closer and closer approximation of the area under the curve. But it would not be possible to add up an infinite number of rectangles, and we also could not multiply by zero for each width, because that would produce a zero sum. Therefore, to find the exact area using the sum of the areas of rectangles, we would have to use the limit process.

We could take the limit of the sums of the areas as the number of rectangles approached infinity. We could also take the limit of the sums as the width of the rectangles approached zero.

The limits of sums can be used to determine the areas under curves, but this is a complex process. Fortunately, a shortcut has evolved to allow us to find the areas under curves without using limits. This shortcut is known as **the Fundamental Theorem of Calculus.**

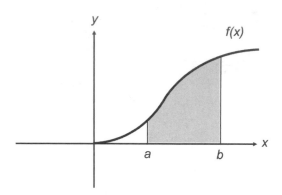

If we wish to find the area under the curve of $f(x)$ between the values of a and b, we can find the definite integral of the curve between those two values. The Fundamental Theorem of Calculus provides us a formula for finding the definite integral:

$$\int_a^b f(x)dx = F(b) - F(a)$$

This formula tells us that to find the definite integral of a function $f(x)$, we evaluate antiderivative F at the value of b. Then we evaluate F at the value of a and subtract.

Another way to phrase the formula is as follows:

$$\int_a^b f'(x)dx = f(b) - f(a)$$

If we are given a derivative function $f'(x)$, the definite integral between a and b equals the original function evaluated at b minus the original function evaluated at a.

The Fundamental Theorem of Calculus allows us to solve definite integral problems relatively quickly. We can just find the antiderivative, then plug in the upper and lower values, and subtract. The resulting number is the area under the curve above the x-axis of the graph. We don't have to use the limits of the sums of the areas of rectangles, which can be a cumbersome process.

Finding definite integrals

To find definite integrals, we take the following steps:

1 Find the indefinite integral.

2 Remove the constant c.

3 Plug in the lower limit value and evaluate the function.

4 Subtract this result from the result of the function evaluated at the upper limit value.

Example

Here is an example of finding a definite integral between the values $x = 2$ and $x = 3$.

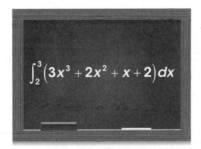

$$\int_2^3 \left(3x^3 + 2x^2 + x + 2\right) dx$$

First, we find the integral, leaving off the constant c:

$$\int_2^3 \left(3x^3 + 2x^2 + x + 2\right) dx = \left[3\frac{x^{3+1}}{(3+1)} + 2\frac{x^{2+1}}{(2+1)} + \frac{x^{1+1}}{(1+1)} + 2\frac{x^{0+1}}{0+1} \right]_2^3$$

$$= \left[\frac{3}{4}x^4 + \frac{2}{3}x^3 + \frac{1}{2}x^2 + 2x \right]_2^3$$

Now we evaluate the result at 2 and subtract it from the result evaluated at 3:

$$\left[\frac{3}{4}x^4 + \frac{2}{3}x^3 + \frac{1}{2}x^2 + 2x \right]_2^3 = \left[\frac{3}{4}(3)^4 + \frac{2}{3}(3)^3 + \frac{1}{2}(3)^2 + 2(3) \right]$$

$$- \left[\frac{3}{4}(2)^4 + \frac{2}{3}(2)^3 + \frac{1}{2}(2)^2 + 2(2) \right]$$

$$= [60.75 + 18 + 4.5 + 6] - [12 + 5.333 + 2 + 4]$$

$$= [89.25] - [23.333]$$

$$\approx 65.92$$

The correct answer is about 65.92.

Using *u*-substitution

Some definite integrals require *u*-substitution to solve. Consider the square root function shown.

$$\int_0^3 \sqrt{x+1}\,dx$$

Always change roots into exponential form.

First, we rewrite the integral into a more convenient form:

$$\int_0^3 \sqrt{x+1}\,dx = \int_0^3 (x+1)^{\frac{1}{2}}\,dx$$

Now we will do a *u*-substitution:

$$u = x + 1$$
$$du = dx$$

Using the substitution, we can find the integral, leaving off the constant *c*:

$$\int_0^3 (x+1)^{\frac{1}{2}}\,dx = \int_{\sim}^{\sim} (u)^{\frac{1}{2}}\,du$$

$$= \left[\frac{u^{\frac{1}{2}+1}}{\left(\frac{1}{2}+1\right)} \right]_{\sim}^{\sim}$$

$$= \left[\frac{u^{\frac{3}{2}}}{\left(\frac{3}{2}\right)} \right]_{\sim}^{\sim}$$

$$= \left[\frac{2}{3} u^{\frac{3}{2}} \right]_{\sim}^{\sim}$$

$$= \left[\frac{2}{3}(x+1)^{\frac{3}{2}} \right]_0^3$$

You may have noticed that we did not write the values at which the integral would be evaluated when we had the expression in terms of u. That is because these would be u-values instead of x-values. We could have solved for the u-values, but there was no point because we were going to change the function back to a function written in terms of x. So, we simply used ~ as a placeholder to indicate that this was a definite integral, and once we had transformed the expression back to one written in terms of x, we put back in the x values.

Finally, we evaluate the result at $x = 0$ and subtract this from the result evaluated at $x = 3$:

$$\left[\frac{2}{3}(x+1)^{\frac{3}{2}}\right]_0^3 = \left[\frac{2}{3}(3+1)^{\frac{3}{2}}\right] - \left[\frac{2}{3}(0+1)^{\frac{3}{2}}\right]$$

$$= \left[\frac{2}{3}(4)^{\frac{3}{2}}\right] - \left[\frac{2}{3}\right]$$

$$\sim 5.33 - \frac{2}{3}$$

$$\approx 4.67$$

The correct answer is about 4.67.

Definite integrals with absolute value

This example shows how to find definite integrals with absolute value.

First, we need to rewrite the integrand to eliminate the absolute value. We examine the integrand and rewrite it as a piecewise function:

$$|x + 2| = \begin{cases} x + 2, & x + 2 \geq 0 \\ -(x + 2), & x + 2 < 0 \end{cases}$$

$$= \begin{cases} x + 2, & x \geq -2 \\ -x - 2, & x < -2 \end{cases}$$

Now we can rewrite the integral in two parts using this information:

$$\int_{-3}^{0} |x + 2| \, dx = \int_{-3}^{-2} (-x - 2) \, dx + \int_{-2}^{0} (x + 2) \, dx$$

Next, we find each integral:

$$\int_{-3}^{-2} (-x - 2) \, dx + \int_{-2}^{0} (x + 2) \, dx = \left[-\frac{1}{2} x^2 - 2x \right]_{-3}^{-2} + \left[\frac{1}{2} x^2 + 2x \right]_{-2}^{0}$$

And we evaluate each:

$$\left[-\frac{1}{2} x^2 - 2x \right]_{-3}^{-2} + \left[\frac{1}{2} x^2 + 2x \right]_{-2}^{0} = \left[-\frac{1}{2}(-2)^2 - 2(-2) \right] - \left[-\frac{1}{2}(-3)^2 - 2(-3) \right] + \left[\frac{1}{2}(0)^2 + 2(0) \right] - \left[\frac{1}{2}(-2)^2 + 2(-2) \right]$$

$$= [-2 + 4] - [-4.5 + 6] + [0] - [2 - 4]$$

$$= 2 - 1.5 + 2$$

$$= 2.5$$

The correct answer is 2.5.

Practice Questions—Finding definite integrals

Directions: Follow the directions given to solve the problems below. You will find the Practice Question Solutions on page 250.

Evaluate the definite integrals.

6. $\int_{1}^{\sqrt{2}} \left(\sqrt{2}\sqrt{x} - 3x^{2} + \frac{2}{x^{2}} \right) dx$

7. $\int_{\frac{\pi}{2}}^{\pi} \cos(x)\, dx$

8. $\int_{0}^{3} |x^{3} - 8|\, dx$

Find the specified area.

9. Find the area of the region within $y = 2x + 5$, the x-axis, the y-axis, and the vertical line $x = 3$.

10. Find the area of the region within $y = \frac{1}{2}x - 2$, the x-axis, the vertical line $x = 4$, and the vertical line $x = 7$.

Chapter Review

Directions: Follow the directions given to solve the problems below. Solutions can be found on page 254.

Find the specified area using integration.

Evaluate the definite integrals.

1. $\int_{-1}^{1}\left(x^2 + 7\right)dx$

2. $\int_{0}^{2}\left(\sqrt{x} - \sqrt{x^3}\right)dx$

3. $\int_{1}^{4}\left(\frac{1}{2}x^3 - 5x + 2\right)dx$

4. $\int_{0}^{\frac{\pi}{4}}7\sec^2(x)dx$

5. $\int_{\frac{\pi}{4}}^{\frac{3\pi}{4}}\csc^2(2x)dx$

6. $\int_{0}^{\frac{\pi}{2}}5\sin^5(x)\cos(x)dx$

7. $\int_{0.6}^{1}\csc^4(x)\cot(x)dx$

8. $\int_{0}^{2}\left(\left|x^3 - 1\right| + 3x\right)dx$

Find the indicated sum.

9. $\sum_{k=1}^{4}\left(7k - 4k^2 + 7\right)$

10. $\sum_{L=1}^{n}\frac{L^3}{(n+1)^2}$, for $n = 3$, 10

Find the indicated area using rectangles.

11. Use four rectangles to find an approximation for the area between the graph of $y = x^{\frac{3}{2}} + 5$, $y = 0$, $x = 1$, and $x = 5$.

12. Find the area of the region within $y = -7x + 6$, the x-axis, the y-axis, and the vertical line $x = 0.8$.

13. Find the area of the region within $y = -x^2 + 4$, the x-axis, the y-axis, and the vertical line $x = 2$.

14. Find the area of the region with $y = -x^3 + 6x - 2$, the horizontal line $y = 1$, the vertical line $x = 0.5$, and the vertical line $x = 2.2$.

15. Find the area of the region between $y = x^2 + 2$ and $y = 2x + 4$.

Practice Question Solutions

Finding area

1. The correct answer is 11.

   To find the sum, we write out each term as k goes from 1 to 3 and add:

   $$\sum_{k=1}^{3}\left(k^2 - k + 1\right) = \left(1^2 - 1 + 1\right) + \left(2^2 - 2 + 1\right) + \left(3^2 - 3 + 1\right)$$
   $$= (1) + (3) + (7)$$
   $$= 11$$

2. The correct answers are $\frac{10}{3}, \frac{95}{6}$, and 80.

   We start by pulling the constant denominator out of the sum:

   $$\sum_{j=1}^{n} \frac{j^2 + j}{n + 1} = \frac{1}{n+1}\sum_{j=1}^{n} j^2 + j$$

   Next, we write this as two sums:

   $$\frac{1}{n+1}\sum_{j=1}^{n} j^2 + j = \frac{1}{n+1}\left(\sum_{j=1}^{n} j^2 + \sum_{j=1}^{n} j\right)$$

   We recall these formulas that will help us here:

   $$\sum_{j=1}^{n} j = \frac{n(n+1)}{2} \text{ and } \sum_{j=1}^{n} j^2 = \frac{n(n+1)(2n+1)}{6}$$

   Applying the formulas, we can write the sums as:

$$\frac{1}{n+1}\left(\sum_{j=1}^{n} j^2 + \sum_{j=1}^{n} j\right) = \frac{1}{n+1}\left(\frac{n(n+1)(2n+1)}{6} + \frac{n(n+1)}{2}\right)$$

$$= \frac{1}{n+1}\left(\frac{n(n+1)(2n+1)}{6} + \frac{3n(n+1)}{6}\right)$$

$$= \frac{1}{n+1}\left(\frac{n(n+1)(2n+1) + 3n(n+1)}{6}\right)$$

$$= \frac{n(2n+1) + 3n}{6}$$

$$= \frac{3n^2 + n + 3n}{6}$$

$$= \frac{3n^2 + 4n}{6}$$

Finally, we can substitute in our values for n to obtain the desired answers:

$$n = 2$$

$$\frac{3(2)^2 + 4(2)}{6} = \frac{12 + 8}{6}$$

$$= \frac{20}{6}$$

$$= \frac{10}{3}$$

$$n = 5$$

$$\frac{3(5)^2 + 4(5)}{6} = \frac{75 + 20}{6}$$

$$= \frac{95}{6}$$

$$n = 12$$

$$\frac{3(12)^2 + 4(12)}{6} = \frac{432 + 48}{6}$$

$$= \frac{480}{6}$$

$$= 80$$

3. The correct answers are $\frac{11}{2}$ and $\frac{3125}{2}$.

First, we write the sum as two sums:

$$\sum_{i=1}^{n}\left(\frac{1}{2}i^3 + 5\right) = \sum_{i=1}^{n}\left(\frac{1}{2}i^3\right) + \sum_{i=1}^{n}(5)$$

For the first sum, we can use the properties of summation to pull the constant outside.

$$\sum_{i=1}^{n}\left(\frac{1}{2}i^3\right) + \sum_{i=1}^{n}(5) = \frac{1}{2}\sum_{i=1}^{n}\left(i^3\right) + \sum_{i=1}^{n}(5)$$

We recall these formulas that will help us here:

$$\sum_{i=1}^{n}\left(i^3\right) = \frac{n^2\left(n+1\right)^2}{4} \text{ and } \sum_{i=1}^{n}(c) = cn$$

Applying the formulas, we can write the sums as:

$$\frac{1}{2}\sum_{i=1}^{n}\left(i^3\right) + \sum_{i=1}^{n}(5) = \left(\frac{1}{2}\right)\left(\frac{n^2\left(n+1\right)^2}{4}\right) + 5n$$

$$= \frac{n^2\left(n+1\right)^2}{8} + \frac{40n}{8}$$

$$= \frac{n^2\left(n^2+2n+1\right)}{8} + \frac{40n}{8}$$

$$= \frac{n^4 + 2n^3 + n^2 + 40n}{8}$$

Finally, we can substitute in our values for n to obtain the desired answers:

$$n = 1$$

$$\frac{(1)^4 + 2(1)^3 + (1)^2 + 40(1)}{8} = \frac{44}{8}$$

$$= \frac{11}{2}$$

$$n = 10$$

$$\frac{(10)^4 + 2(10)^3 + (10)^2 + 40(10)}{8} = \frac{3125}{2}$$

4. The correct answer is 45.

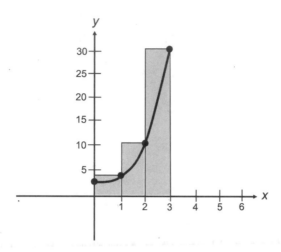

Note: Figure not drawn to scale.

To understand this problem, we graph the function between $x = 0$ and $x = 3$, as shown in the figure above. We then divide the region into three different sections, each with a width of 1, and create a rectangle in each section. We can find the height of each rectangle by realizing that the rectangles we have drawn have their highest points at their right edges. We can solve the function at each of these edges to find the height of each of these rectangles.

$$f(1) = (1)^3 + 3$$
$$= 4$$

$$f(2) = (2)^3 + 3$$
$$= 11$$

$$f(3) = (3)^3 + 3$$
$$= 30$$

Now we can find the approximation for the area by finding the area of each rectangle and adding them together:

$$\text{Area} = (4)(1) + (11)(1) + (30)(1)$$
$$= 45$$

We can see from graphing the function and the rectangles on the same graph that in this case, the approximation overestimates the area. Other approximations are possible. We could have set the height of each rectangle to be the same as the height of the function at the left-most edges of the rectangles. That approximation would have underestimated the area. A

Third possibility would have been to set the height of each rectangle such that the midpoints of the rectangles were values on the function. This would have been the best approximation of the area out of the three possibilities mentioned.

5. The correct answer is 6.14.

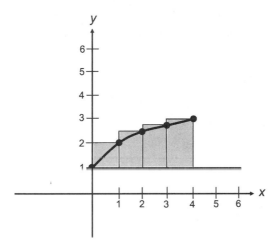

To understand this problem, we graph the function between $x = 0$ and $x = 4$, as shown in the figure. We then divide the region into four different sections, each with a width of 1, and create a rectangle in each section. We can find the height of each rectangle by realizing that the rectangles we have drawn have their highest points at their right edges. We can solve the function at each of these edges and then subtract 1 to find the height of each of these rectangles.

$$f(1) - 1 = \sqrt{1} + 1 - 1$$
$$= 1$$

$$f(2) - 1 = \sqrt{2} + 1 - 1$$
$$\approx 1.41$$

$$f(3) - 1 = \sqrt{3} + 1 - 1$$
$$\approx 1.73$$

$$f(4) - 1 = \sqrt{4} + 1 - 1$$
$$= 2$$

Now we can find the approximation for the area by finding the area of each rectangle and adding them together:

$$\text{Area} \approx (1)(1) + (1.41)(1) + (1.73)(1) + (2)(1)$$
$$\approx 6.14$$

We can see from graphing the function and the rectangles on the same graph that in this case, the approximation overestimates the area. Other approximations are possible.

Finding definite integrals

6.

The correct answer is –0.6.

First, we rewrite the integral into a more convenient form:

$$\int_1^{\sqrt{2}} \left(\sqrt{2}\sqrt{x} - 3x^2 + \frac{2}{x^2} \right) dx = \int_1^{\sqrt{2}} \left(\sqrt{2}x^{\frac{1}{2}} - 3x^2 + 2x^{-2} \right) dx$$

Now we find the integral:

$$\int_1^{\sqrt{2}} \left(\sqrt{2}x^{\frac{1}{2}} - 3x^2 + 2x^{-2} \right) dx = \left[\sqrt{2}\,\frac{x^{\frac{1}{2}+1}}{\left(\frac{1}{2}+1\right)} - 3\frac{x^{2+1}}{(2+1)} + 2\frac{x^{-2+1}}{(-2+1)} \right]_1^{\sqrt{2}}$$

$$= \left[\frac{2\sqrt{2}}{3}x^{\frac{3}{2}} - x^3 - \frac{2}{x} \right]_1^{\sqrt{2}}$$

Next, we evaluate the result at 1 and subtract it from the result evaluated at $\sqrt{2}$:

$$\left[\frac{2\sqrt{2}}{3}x^{\frac{3}{2}} - x^3 - \frac{2}{x}\right]_1^{\sqrt{2}} = \left[\frac{2\sqrt{2}}{3}\left(\sqrt{2}\right)^{\frac{3}{2}} - \left(\sqrt{2}\right)^3 - \frac{2}{\left(\sqrt{2}\right)}\right] - \left[\frac{2\sqrt{2}}{3}(1)^{\frac{3}{2}} - (1)^3 - \frac{2}{(1)}\right]$$

$$= \left[\frac{2\sqrt{2}}{3}\left(2^{\frac{1}{2}}\right)^{\frac{3}{2}} - \left(2^{\frac{1}{2}}\right)^3 - \frac{2}{\sqrt{2}}\frac{\sqrt{2}}{\sqrt{2}}\right] - \left[\frac{2\sqrt{2}}{3} - 1 - 2\right]$$

$$= \left[\frac{2\sqrt{2}}{3}2^{\frac{3}{4}} - 2^{\frac{3}{2}} - \frac{2\sqrt{2}}{2}\right] - \left[\frac{2\sqrt{2}}{3} - \frac{9}{3}\right]$$

$$= \left[\frac{2^{\frac{7}{4}}\sqrt{2}}{3} - 2^{\frac{3}{2}} - \sqrt{2}\right] - \left[\frac{2\sqrt{2} - 9}{3}\right]$$

$$\approx [-2.657] - [-2.057]$$

$$\approx -0.6$$

7. The correct answer is –1.

First, we find the integral:

$$\int_{\frac{\pi}{2}}^{\pi} \cos(x)\,dx = \left[\sin(x)\right]_{\frac{\pi}{2}}^{\pi}$$

Now we evaluate the result at $\frac{\pi}{2}$ and subtract it from the result evaluated at π:

$$\left[\sin(x)\right]_{\frac{\pi}{2}}^{\pi} = \left[\sin(\pi)\right] - \left[\sin\left(\frac{\pi}{2}\right)\right]$$

Using the unit circle, we can evaluate these sine functions and get an answer:

$$\left[\sin(\pi)\right] - \left[\sin\left(\frac{\pi}{2}\right)\right] = [0] - [1]$$

$$= -1$$

8. The correct answer is 20.25.

First, we need to rewrite the integrand to eliminate the absolute value. We examine the integrand and rewrite it as a piecewise function:

$$\left|x^3 - 8\right| = \begin{cases} x^3 - 8, & x^3 - 8 \geq 0 \\ -\left(x^3 - 8\right), & x^3 - 8 < 0 \end{cases}$$

$$= \begin{cases} x^3 - 8, & x^3 \geq 8 \\ -x^3 + 8, & x^3 < 8 \end{cases}$$

$$= \begin{cases} x^3 - 8, & x \geq 2 \\ -x^3 + 8, & x < 2 \end{cases}$$

Now we can rewrite the integral in two parts using this information:

$$\int_0^3 \left|x^3 - 8\right| dx = \int_0^2 \left(-x^3 + 8\right) dx + \int_2^3 \left(x^3 - 8\right) dx$$

Next, we find each integral:

$$\int_0^2 \left(-x^3 + 8\right) dx + \int_2^3 \left(x^3 - 8\right) dx = \left[-\frac{1}{4}x^4 + 8x\right]_0^2 + \left[\frac{1}{4}x^4 - 8x\right]_2^3$$

Finally, we evaluate each:

$$\left[-\frac{1}{4}x^4 + 8x\right]_0^2 + \left[\frac{1}{4}x^4 - 8x\right]_2^3 = \left[-\frac{1}{4}(2)^4 + 8(2)\right] - \left[-\frac{1}{4}(0)^4 + 8(0)\right] + \left[\frac{1}{4}(3)^4 - 8(3)\right] - \left[\frac{1}{4}(2)^4 - 8(2)\right]$$

$$= [-4 + 16] - [0] + [20.25 - 24] - [4 - 16]$$
$$= [12] + [-3.75] - [-12]$$
$$= 20.25$$

9. The correct answer is 33.

The area we seek is the area under the line $y = 2x + 5$, between $x = 0$ and $x = 3$. To find this area, we set up an integral:

$$\text{Area} = \int_0^3 (2x + 5) dx$$

Now we solve the integral:

$$\int_0^3 (2x + 5) dx = \left[\frac{2}{3}x^3 + 5x\right]_0^3$$

Finally, we evaluate the integral between $x = 0$ and $x = 3$:

$$\left[\frac{2}{3}x^3 + 5x\right]_0^3 = \left[\frac{2}{3}(3)^3 + 5(3)\right] - \left[\frac{2}{3}(0)^3 + 5(0)\right]$$
$$= [18 + 15] - 0$$
$$= 33$$

10. The correct answer is 2.25.

   The area we seek is the area under the line $y = \frac{1}{2}x - 2$ between $x = 4$ and $x = 7$. To find this area, we set up an integral:

   $$Area = \int_4^7 \left(\frac{1}{2}x - 2\right)dx$$

Now we solve the integral.

$$\int_4^7 \left(\frac{1}{2}x - 2\right)dx = \left[\frac{1}{4}x^2 - 2x\right]_4^7$$

Finally, we evaluate the integral between $x = 4$ and $x = 7$:

$$\left[\frac{1}{4}x^2 - 2x\right]_4^7 = \left[\frac{1}{4}(7)^2 - 2(7)\right] - \left[\frac{1}{4}(4)^2 - 2(4)\right]$$
$$= [12.25 - 14] - [4 - 8]$$
$$= [-1.75] - [-4]$$
$$= 2.25$$

Chapter Review Solutions

1. The correct answer is 14.67.

First, we find the integral.

$$\int_{-1}^{1}\left(x^2 + 7\right)dx = \left[\frac{x^{2+1}}{(2+1)} + 7\frac{x^{0+1}}{(0+1)}\right]_{-1}^{1}$$

$$= \left[\frac{1}{3}x^3 + 7x\right]_{-1}^{1}$$

Now we evaluate the result at –1 and subtract it from the result evaluated at 1:

$$\left[\frac{1}{3}x^3 + 7x\right]_{-1}^{1} = \left[\frac{1}{3}(1)^3 + 7(1)\right] - \left[\frac{1}{3}(-1)^3 + 7(-1)\right]$$

$$= \left[\frac{1}{3} + 7\right] - \left[-\frac{1}{3} - 7\right]$$

$$= \frac{1}{3} + 7 + \frac{1}{3} + 7$$

$$\approx 14.67$$

2. The correct answer is –0.377.

First, we rewrite the integral into a more convenient form:

$$\int_{0}^{2}\left(\sqrt{x} - \sqrt{x^3}\right)dx = \int_{0}^{2}\left(x^{\frac{1}{2}} - \left(x^{\frac{1}{2}}\right)^3\right)dx$$

$$= \int_{0}^{2}\left(x^{\frac{1}{2}} - x^{\frac{3}{2}}\right)dx$$

Now we find the integral:

$$\int_0^2 \left(x^{\frac{1}{2}} - x^{\frac{3}{2}} \right) dx = \left[\frac{x^{\frac{1}{2}+1}}{\left(\frac{1}{2}+1\right)} - \frac{x^{\frac{3}{2}+1}}{\left(\frac{3}{2}+1\right)} \right]_0^2$$

$$= \left[\frac{x^{\frac{3}{2}}}{\left(\frac{3}{2}\right)} - \frac{x^{\frac{5}{2}}}{\left(\frac{5}{2}\right)} \right]_0^2$$

$$= \left[\frac{2}{3} x^{\frac{3}{2}} - \frac{2}{5} x^{\frac{5}{2}} \right]_0^2$$

Finally, we evaluate the result at 0 and subtract it from the result evaluated at 2:

$$\left[\frac{2}{3} x^{\frac{3}{2}} - \frac{2}{5} x^{\frac{5}{2}} \right]_0^2 = \left[\frac{2}{3}(2)^{\frac{3}{2}} - \frac{2}{5}(2)^{\frac{5}{2}} \right] - \left[\frac{2}{3}(0)^{\frac{3}{2}} - \frac{2}{5}(0)^{\frac{5}{2}} \right]$$

$$\approx [1.886 - 2.263] - [0]$$

$$\approx -0.377$$

Here the answer is negative, because if we were to observe the graph of the function, we would see that there is more area under the x-axis than above the x-axis between x = 0 and x = 2.

3. The correct answer is $\frac{3}{8}$.

First, we find the integral:

$$\int_1^4 \left(\frac{1}{2} x^3 - 5x + 2 \right) dx = \left[\left(\frac{1}{2}\right) \frac{x^{3+1}}{(3+1)} - 5 \frac{x^{1+1}}{(1+1)} + 2 \frac{x^{0+1}}{(0+1)} \right]_1^4$$

$$= \left[\frac{1}{8} x^4 - \frac{5}{2} x^2 + 2x \right]_1^4$$

Now we evaluate the result at 1 and subtract it from the result evaluated at 4:

$$\left[\frac{1}{8}x^4 - \frac{5}{2}x^2 + 2x\right]_1^4 = \left[\frac{1}{8}(4)^4 - \frac{5}{2}(4)^2 + 2(4)\right] - \left[\frac{1}{8}(1)^4 - \frac{5}{2}(1)^2 + 2(1)\right]$$

$$= [32 - 40 + 8] - \left[\frac{1}{8} - \frac{5}{2} + 2\right]$$

$$= [0] - \left[\frac{1}{8} - \frac{20}{8} + \frac{16}{8}\right]$$

$$= -\left[-\frac{3}{8}\right]$$

$$= \frac{3}{8}$$

4.

The correct answer is 7.

First, we can pull the 7 outside the integrand since it is a constant:

$$\int_0^{\frac{\pi}{4}} 7\sec^2(x)\,dx = 7\int_0^{\frac{\pi}{4}} \sec^2(x)\,dx$$

Now we recognize that the derivative of $\tan(x)$ is $\sec^2(x)$. Integration is the inverse of differentiation, so going the other way, the integral of $\sec^2(x)$ will equal $\tan(x)$.

$$7\int_0^{\frac{\pi}{4}} \sec^2(x)\,dx = 7\left[\tan(x)\right]_0^{\frac{\pi}{4}}$$

Using the unit circle, we evaluate the tangent at 0 and subtract that from the tangent evaluated at $\frac{\pi}{4}$:

$$7\left[\tan(x)\right]_{0}^{\frac{\pi}{4}} = 7\left[\tan\left(\frac{\pi}{4}\right) - \tan(0)\right]$$
$$= 7\left[1 - 0\right]$$
$$= 7$$

5. The correct answer is 0.

We start with a *u*-substitution:

$$u = 2x$$
$$du = 2dx$$
$$\frac{1}{2}du = dx$$

We recall that the derivative of $\cot(x)$ is $-\csc^2(x)$. Going the other direction, the integral of $-\csc^2(x)$ is $\cot(x)$. The negative, which is just a constant, can be moved to either side. Using this knowledge and the *u*-substitution, we can find the integral:

$$\int_{\frac{\pi}{4}}^{\frac{3\pi}{4}} \csc^2(2x)\,dx = \frac{1}{2}\int_{\sim}^{\sim} \csc^2(u)\,du$$
$$= \frac{1}{2}\left[-\cot(u)\right]_{\sim}^{\sim}$$
$$= \frac{1}{2}\left[-\cot(2x)\right]_{\frac{\pi}{4}}^{\frac{3\pi}{4}}$$

We can evaluate this at the two given values and subtract:

$$\frac{1}{2}\left[-\cot(2x)\right]_{\frac{\pi}{4}}^{\frac{3\pi}{4}} = \frac{1}{2}\left[-\cot\left((2)\frac{3\pi}{4}\right) - \left(-\cot\left((2)\frac{\pi}{4}\right)\right)\right]$$
$$= \frac{1}{2}\left[-\cot\left(\frac{3\pi}{2}\right) - \left(-\cot\left(\frac{\pi}{2}\right)\right)\right]$$
$$= \frac{1}{2}\left[-\frac{\cos\left(\frac{3\pi}{2}\right)}{\sin\left(\frac{3\pi}{2}\right)} - \left(-\frac{\cos\left(\frac{\pi}{2}\right)}{\sin\left(\frac{\pi}{2}\right)}\right)\right]$$
$$= \frac{1}{2}\left[-\frac{0}{1} - \left(-\frac{0}{-1}\right)\right]$$
$$= 0$$

6. The correct answer is $\frac{5}{6}$.

We can see that if we let $u = \sin(x)$, the derivative $\cos(x)$ is present:

$$u = \sin(x)$$
$$du = \cos(x)\,dx$$

Using the substitution, we can find the integral:

$$\int_0^{\frac{\pi}{2}} 5\sin^5(x)\cos(x)\,dx = 5\int_{\sim} u^5\,du$$
$$= 5\left[\frac{1}{6}u^6\right]_{\sim}^{\sim}$$
$$= \frac{5}{6}\left[\sin^6(x)\right]_0^{\frac{\pi}{2}}$$

We have pulled the 5 and the $\frac{1}{6}$ out, since they are both constants. Now we evaluate this at the two given values and subtract:

$$\frac{5}{6}\left[\sin^6(x)\right]_0^{\frac{\pi}{2}} = \frac{5}{6}\left[\sin^6\left(\frac{\pi}{2}\right) - \sin^6(0)\right]$$
$$= \frac{5}{6}\left[\left(\sin\left(\frac{\pi}{2}\right)\right)^6 - (\sin(0))^6\right]$$
$$= \frac{5}{6}\left[(1)^6 - (0)^6\right]$$
$$= \frac{5}{6}$$

7. The correct answer is 1.96.

We can see that if we let $u = \csc(x)$, the derivative $\csc(x)\cot(x)$ is present.

$$u = \csc(x)$$
$$du = -\csc(x)\cot(x)\,dx$$
$$-du = \csc(x)\cot(x)\,dx$$

We use u-substitution to find the integral:

$$\int_{0.6}^{1} \csc^4(x)\cot(x)\,dx = \int_{0.6}^{1} \csc^3(x)\csc(x)\cot(x)\,dx$$

$$= -\int_{\sim}^{\sim} u^3\,du$$

$$= -\left[\frac{1}{4}u^4\right]_{\sim}^{\sim}$$

$$= -\left[\frac{1}{4}\csc^4(x)\right]_{0.6}^{1}$$

$$= -\frac{1}{4}\left[\csc^4(x)\right]_{0.6}^{1}$$

Now we evaluate the result at 0.6 and subtract it from the result evaluated at 1:

$$-\frac{1}{4}\left[\csc^4(x)\right]_{0.6}^{1} = -\frac{1}{4}\left[\csc^4(1) - \csc^4(0.6)\right]$$

$$= -\frac{1}{4}\left[\left(\frac{1}{\sin(1)}\right)^4 - \left(\frac{1}{\sin(0.6)}\right)^4\right]$$

$$= -\frac{1}{4}[1.99 - 9.84]$$

$$= 1.96$$

8. The correct answer is 9.5.

First, we need to rewrite the integrand to eliminate the absolute value. We examine the absolute value part of the integrand and rewrite it as a piecewise function:

$$\left|x^3 - 1\right| = \begin{cases} x^3 - 1, & x^3 - 1 \geq 0 \\ -(x^3 - 1), & x^3 - 1 < 0 \end{cases}$$

$$= \begin{cases} x^3 - 1, & x^3 \geq 1 \\ -x^3 + 1, & x^3 < 1 \end{cases}$$

$$= \begin{cases} x^3 - 1, & x \geq 1 \\ -x^3 + 1, & x < 1 \end{cases}$$

Now we can rewrite the integral in two parts using this information:

$$\int_{0}^{2}\left(\left|x^3 - 1\right| + 3x\right)dx = \int_{0}^{1}\left(-x^3 + 1 + 3x\right)dx + \int_{1}^{2}\left(x^3 - 1 + 3x\right)dx$$

$$= \int_{0}^{1}\left(-x^3 + 3x + 1\right)dx + \int_{1}^{2}\left(x^3 + 3x - 1\right)dx$$

Next, we find each integral:

$$\int_0^1 \left(-x^3 + 3x + 1\right) dx + \int_1^2 \left(x^3 + 3x - 1\right) dx = \left[-\frac{1}{4}x^4 + \frac{3}{2}x^2 + x\right]_0^1 + \left[\frac{1}{4}x^4 + \frac{3}{2}x^2 - x\right]_1^2$$

Finally, we evaluate each:

$$\left[-\frac{1}{4}x^4 + \frac{3}{2}x^2 + x\right]_0^1 + \left[\frac{1}{4}x^4 + \frac{3}{2}x^2 - x\right]_1^2 = \left[-\frac{1}{4}(1)^4 + \frac{3}{2}(1)^2 + (1)\right] - \left[-\frac{1}{4}(0)^4 + \frac{3}{2}(0)^2 + (0)\right]$$

$$+ \left[\frac{1}{4}(2)^4 + \frac{3}{2}(2)^2 - (2)\right] - \left[\frac{1}{4}(1)^4 + \frac{3}{2}(1)^2 - (1)\right]$$

$$= \left[-\frac{1}{4} + \frac{3}{2} + 1\right] - [0] + [4 + 6 - 2] - \left[\frac{1}{4} + \frac{3}{2} - 1\right]$$

$$= \left[\frac{9}{4}\right] + [8] - \left[\frac{3}{4}\right]$$

$$= 9.5$$

9.

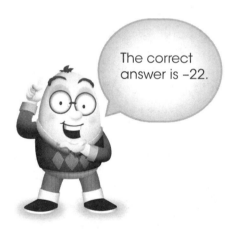

The correct answer is –22.

To find the sum, we write out each term as k goes from 1 to 4 and add:

$$\sum_{k=1}^{4} \left(7k - 4k^2 + 7\right) = \left(7(1) - 4(1)^2 + 7\right) + \left(7(2) - 4(2)^2 + 7\right) + \left(7(3) - 4(3)^2 + 7\right) + \left(7(4) - 4(4)^2 + 7\right)$$

$$= 10 + 5 - 8 - 29$$

$$= -22$$

10. The correct answers are 2.25 and 25.

We start by pulling the constant denominator out of the sum:

$$\sum_{L=1}^{n} \frac{L^3}{(n+1)^2} = \frac{1}{(n+1)^2} \sum_{L=1}^{n} L^3$$

We recall this formula that will help us here:

$$\sum_{i=1}^{n} i^3 = \frac{n^2(n+1)^2}{4}$$

Applying the formula, we can write the sum as:

$$\frac{1}{(n+1)^2} \sum_{L=1}^{n} L^3 - \frac{1}{(n+1)^2} \frac{n^2(n+1)^2}{4}$$

$$= \frac{n^2}{4}$$

Finally, we can substitute in our values for n to obtain the desired answers:

$$n = 3$$
$$\frac{(3)^2}{4} = \frac{9}{4}$$
$$= 2.25$$

$$n = 10$$
$$\frac{(10)^2}{4} = \frac{100}{4}$$
$$= 25$$

11. The correct answer is 37.024.

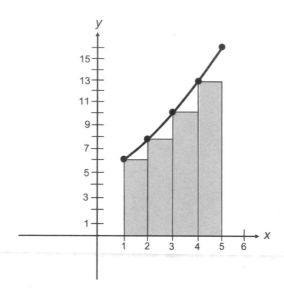

Note: Figure not drawn to scale.

To understand this problem, we graph the function between $x = 1$ and $x = 5$, as shown in the figure. We then divide the region into four different sections, each with a width of 1, and create a rectangle in each section. We can find the height of each rectangle by realizing that the rectangles we have drawn have their highest points at their left edges. We can solve the function at each of these edges to find the height of each rectangle.

$$f(1) = (1)^{\frac{3}{2}} + 5$$
$$= 1 + 5$$
$$= 6$$

$$f(2) = (2)^{\frac{3}{2}} + 5$$
$$\approx 7.828$$

$$f(3) = (3)^{\frac{3}{2}} + 5$$
$$\approx 10.196$$

$$f(4) = (4)^{\frac{3}{2}} + 5$$
$$= 13$$

Now we can find the approximation for the area by finding the area of each rectangle and adding them together:

$$\text{Area} \approx (1)(6) + (1)(7.828) + (1)(10.196) + (1)(13)$$
$$\approx 6 + 7.828 + 10.196 + 13$$
$$\approx 37.024$$

We can see from graphing the function and the rectangles on the same graph that in this case, the approximation underestimates the area. Other approximations are possible.

12. The correct answer is 2.56.

The area we seek is the area under the line $y = -7x + 6$ between $x = 0$ and $x = 0.8$. To find this area, we set up an integral:

$$\text{Area} = \int_0^{0.8} (-7x + 6)\,dx$$

Now we solve the integral:

$$\int_0^{0.8} (-7x + 6)\,dx = \left[-\frac{7}{2}x^2 + 6x\right]_0^{0.8}$$

Finally, we evaluate the integral between $x = 0$ and $x = 0.8$:

$$\left[-\frac{7}{2}x^2 + 6x\right]_0^{0.8} = \left[-\frac{7}{2}(0.8)^2 + 6(0.8)\right] - \left[-\frac{7}{2}(0)^2 + 6(0)\right]$$
$$= [-2.24 + 4.8] - [0]$$
$$= 2.56$$

13. The correct answer is 5.33.

The area we seek is the area under the curve $y = -x^2 + 4$, between $x = 0$ and $x = 2$. To find this area, we set up an integral:

$$\text{Area} = \int_0^2 (-x^2 + 4)\,dx$$

Now we solve the integral:

$$\int_0^2 (-x^2 + 4)\,dx = \left[-\frac{1}{3}x^3 + 4x\right]_0^2$$

Finally, we evaluate the integral between $x = 0$ and $x = 2$:

$$\left[-\frac{1}{3}x^3 + 4x\right]_0^2 = \left[-\frac{1}{3}(2)^3 + 4(2)\right] - \left[-\frac{1}{3}(0)^3 + 4(0)\right]$$
$$= [-2.67 + 8] - [0]$$
$$= 5.33$$

14. The correct answer is 2.826.

   The area we seek is the area between the curve $y = -x^3 + 6x - 2$ and the horizontal line $y = 1$, between $x = 0.5$ and $x = 2.2$. To find this area, we set up a difference of two integrals:

   $$\text{Area} = \int_{0.5}^{2.2}\left(-x^3 + 6x - 2\right)dx - \int_{0.5}^{2.2}(1)\,dx$$

   Now we solve the integrals:

   $$\int_{0.5}^{2.2}\left(-x^3 + 6x - 2\right)dx - \int_{0.5}^{2.2}(1)\,dx = \left[-\frac{1}{4}x^4 + 3x^2 - 2x\right]_{0.5}^{2.2} - \left[x\right]_{0.5}^{2.2}$$

   Finally, we evaluate the integrals between $x = 0.5$ and $x = 2.2$:

   $$\left[-\frac{1}{4}x^4 + 3x^2 - 2x\right]_{0.5}^{2.2} - \left[x\right]_{0.5}^{2.2} = \left[-\frac{1}{4}(2.2)^4 + 3(2.2)^2 - 2(2.2)\right] - \left[-\frac{1}{4}(0.5)^4 + 3(0.5)^2 - 2(0.5)\right] - \left[[2.2] - [0.5]\right]$$
   $$\approx [-5.86 + 14.52 - 4.4] - [-0.016 + 0.75 - 1] - [1.7]$$
   $$\approx 4.26 + 0.266 - 1.7$$
   $$\approx 2.826$$

15. The correct answer is 6.936.

   In order to find this area, we need to find where the two graphs cross each other. These will be the limits of our integration (we will find the area between the two graphs in this region). To find where the two graphs overlap, we set the two equations equal to each other and solve for x:

$$x^2 + 2 = 2x + 4$$
$$x^2 - 2x - 2 = 0$$

$$x = \frac{-(-2) \pm \sqrt{(-2)^2 - 4(1)(-2)}}{2(1)}$$

$$= \frac{2 \pm \sqrt{4 + 8}}{2}$$

$$= \frac{2 \pm \sqrt{12}}{2}$$

$$= \frac{2 \pm \sqrt{4}\sqrt{3}}{2}$$

$$= \frac{2 \pm 2\sqrt{3}}{2}$$

$$= 1 \pm \sqrt{3}$$

$$= 1 + \sqrt{3} \approx 2.73,\ 1 - \sqrt{3} \approx -0.732$$

Now we can set up an integral to represent the area between the curves. The top curve is $y = 2x + 4$, and the bottom curve is $y = x^2 + 2$. We will integrate between $1 - \sqrt{3}$ and $1 + \sqrt{3}$:

$$\text{Area} = \int_{1-\sqrt{3}}^{1+\sqrt{3}} (2x + 4)\, dx - \int_{1-\sqrt{3}}^{1+\sqrt{3}} \left(x^2 + 2\right) dx$$

First, we find the integrals:

$$\int_{1-\sqrt{3}}^{1+\sqrt{3}} (2x + 4)\, dx - \int_{1-\sqrt{3}}^{1+\sqrt{3}} \left(x^2 + 2\right) dx = \left[x^2 + 4x \right]_{1-\sqrt{3}}^{1+\sqrt{3}} - \left[\frac{1}{3}x^3 + 2x \right]_{1-\sqrt{3}}^{1+\sqrt{3}}$$

Then we evaluate the integrals between $1 - \sqrt{3}$ and $1 + \sqrt{3}$:

$$\left[x^2 + 4x \right]_{1-\sqrt{3}}^{1+\sqrt{3}} - \left[\frac{1}{3}x^3 + 2x \right]_{1-\sqrt{3}}^{1+\sqrt{3}} = \left[\left(1 + \sqrt{3}\right)^2 + 4\left(1 + \sqrt{3}\right) \right] - \left[\left(1 - \sqrt{3}\right)^2 + 4\left(1 - \sqrt{3}\right) \right]$$

$$- \left[\left[\frac{1}{3}\left(1 + \sqrt{3}\right)^3 + 2\left(1 + \sqrt{3}\right) \right] - \left[\frac{1}{3}\left(1 - \sqrt{3}\right)^3 + 2\left(1 - \sqrt{3}\right) \right] \right]$$

$$\approx [7.46 + 10.93] - [0.536 - 2.93] - \left[[6.797 + 5.46] - [-0.131 - 1.46] \right]$$

$$\approx [18.39] - [-2.394] - \left[[12.257] - [-1.591] \right]$$

$$\approx 18.39 + 2.394 - [12.257 + 1.591]$$

$$\approx 20.784 - 13.848$$

$$\approx 6.936$$

Remember you can visit www.petersonspublishing.com for additional calculus practice exercises.

NOTES

NOTES

NOTES

NOTES

NOTES

NOTES

NOTES

NOTES

NOTES

NOTES